AF556103

Research on Plant Tissue Culture

Research on Plant Tissue Culture

Dr. Sachin Kumar

RANDOM PUBLICATIONS
NEW DELHI (INDIA)

Research on Plant Tissue Culture

ISBN 978-93-5111-219-8

Published in 2014 in India by

RANDOM PUBLICATIONS

4376-A/4B, Gali Murari Lal, Ansari Road
New Delhi-110 002
Phone : +91-11-43580356, +91-11-23289044
e-mail: randomexports@gmail.com, sales@randompublications.com, info@randompublications.com

Reprinted 2021

Type Setting by : Keystoneprintads, Delhi-110051
Digitally Printed at: Replika Press Pvt. Ltd.

Preface

The Plant Tissue Culture laboratory is constantly trying to achieve as much as it possibly can. There are numerous research projects which are undertaken and at the same time, the production unit also has numerous projects for the large scale production of saplings of numerous plants. Another purpose for which plant tissue culture is uniquely suited is in the obtaining, maintaining, and mass propagating of specific pathogen-free plants. The concept behind indexing plants free of pests is closely allied to the concept of using tissue culture as a selection system. Plant tissues known to be free of the pathogen under consideration (viral, bacterial, or fungal) are physically selected as the explant for tissue culture. In most cases, the apical domes of rapidly elongating shoot tips are chosen.

Generalisations about plant growth regulators and their use in plant cell culture media have been developed from initial observations made in the 1950s. There is, however, some considerable difficulty in predicting the effects of plant growth regulators: this is because of the great differences in culture response between species, cultivars and even plants of the same cultivar grown under different conditions. However, some principles do hold true and have become the paradigm on which most plant tissue culture regimes are based. Auxins and cytokinins are the most widely used plant growth regulators in plant tissue culture and are usually used together, the ratio of the auxin to the cytokinin determining the type of culture established or regenerated. A high auxin to cytokinin ratio generally favours root formation, whereas a high cytokinin to auxin ratio favours shoot formation. An intermediate ratio favours callus production.

Tissue engineering has evolved from the use of biomaterials for bone substitution that fulfill the clinical demands of biocompatibility, biodegradability, non-immunogeneity, structural strength and porosity. Porous scaffolds have been developed in many forms and materials, but few reached the need of adequate physical, biological and mechanical properties.

In the present paper we report the preparation of hybrid porous polyvinyl alcohol (PVA)/bioactive glass through the sol-gel route, using partially and fully hydrolyzed polyvinyl alcohol, and perform structural characterization. This book on plant tissue culture deals with the principles and most of the techniques being employed in the tissue culture.

I thank all members of my team who have helped in the preparation of the book. My special thanks go to "Random Publications" who have published the book.

– Dr. Sachin Kumar

Contents

Preface .... *v-vi*

1. Plant Growth Regulators and Tissue Culture 1

Demonstration of Tissue Culture for Teaching 2
Regulatory Mutations and There Behaviour 4
Basic Steps of Plant Tissue Culture 8
Theory of Tissue Culture 11
Plant Tissue Culture 13
Plant Tissue Culture Requirement 16

2. Plant Improvement through Tissue Culture 19

Plant Improvement in Cell Walls 19
Selection of Plants With Enhanced Stress or Pest Resistance 25
Improving Techniques of the Quality System 27
Tissue Culture of Lilacs 29
Vitro Flowering of Rose 29
Organogenesis 30
Direct Shoot Organogenesis From Leaf Explants of Chrysanthemum 32
Direct and Indirect Shoot Organogenesis 33
Shoot Organogenesis from Petunia Leaves 33
Non-zygotic Embryogenesis 33
Somatic Embryogenesis from Seeds of Melon 35
Improvement of Nutrient Quality 36

3. Cell Culture of Plants: Methods and Techniques 43

Methods and Techniques 43
Development of Plant Cell 46
Cell Culture 52
The Production of Cells 53
Structure of Cell Wall 59
Methods of Single Cell Isolation 64

4. Research on Plant Movements and Development 71

Plant Movements 71
Plant Growth in Different Ways 76
Research on Plant Development 77
Transition and Vegetative of Plant Reproductive Development 80

5. **Gene Transformation Methods in Plants** 87

Transformation of Plant Genome 87
Process of Gene Transfer in Plants 96
Method of Plant Transformation 101
Vector-mediated or Indirect Gene Transfer 103
Significance of Plants Cell Expansion 104
Vectorless or Direct Gene Transfer 108

6. **Plant Cells Model and its Parts** 110

Plant Cells 110
Functions of Cell Walls in Plant 113
Cell Model of Plant 125
Structure and Function of Plants 127

7. **Techniques for Cloning Plants** 138

Method of DNA Cloning and Manipulate DNA 140
Cloning Unicellular Organisms 143
Plant Gene Cloning 145
Producing Clones: Plant Life 154

8. **Composite Scaffolds for Bone Tissue Engineering** 156

Hybrid Biomaterials for Bone Tissue Engineerings 157
Tissue Engineering in Combination of Cell 160
Precursors of Tissue Engineering 168
Particulate Bioceramics for Composite Scaffolds 179
Bone Tissue Engineering Using Foetal Cell Therapys 182
Tissue Engineering in Cardiovascular Surgery 185
Preparation of The Pva-bioactive Glass Hybrid Foams 188
The Tissue Engineers as a Group 194

9. **Research on Bamboo Micropropagation** 209

Flowering of Bamboo in Tissue Culture 210
Commercial Micropropagation of Bamboo 211
Molecular Basis of Somatic Mutations in Bamboo 214

10. **Genetic Engineering in Plant Breeding** 215

Aim of Plant Breeding 215
Conventional Breeding 221
Scope of Plant Breeding 227
Plant Genetics 230
Genetic Engineering 235
Biological Energy-ADP and ATP Respiration 238
Genetic Engineering Technology 240
Gene Mapping 249

11. **Tissue Modifications Caused by Parastites** 263

The Mechanism of Feeding by Parasites 274
The Fungi Enzymes 276

Index 279

1

Plant Growth Regulators and Tissue Culture

Generalisations about plant growth regulators and their use in plant cell culture media have been developed from initial observations made in the 1950s. There is, however, some considerable difficulty in predicting the effects of plant growth regulators: this is because of the great differences in culture response between species, cultivars and even plants of the same cultivar grown under different conditions. However, some principles do hold true and have become the paradigm on which most plant tissue culture regimes are based. Auxins and cytokinins are the most widely used plant growth regulators in plant tissue culture and are usually used together, the ratio of the auxin to the cytokinin determining the type of culture established or regenerated. A high auxin to cytokinin ratio generally favours root formation, whereas a high cytokinin to auxin ratio favours shoot formation. An intermediate ratio favours callus production.

CULTURE TYPES

Cultures are generally initiated from sterile pieces of a whole plant. These pieces are termed 'explants', and may consist of pieces of organs, such as leaves or roots, or may be specific cell types, such as pollen or endosperm. Many features of the explant are known to affect the efficiency of culture initiation. Generally, younger, more rapidly growing tissue (or tissue at an early stage of development) is most effective. Several different culture types most commonly used in plant transformation studies will now be examined in more detail.

Callus

Explants, when cultured on the appropriate medium, usually with both an auxin and a cytokinin, can give rise to an unorganised, growing and dividing mass of cells. It is thought that any plant tissue can be used as an explant, if the correct conditions are found. In culture, this proliferation can be maintained more or less indefinitely, provided that the callus is subcultured on to fresh medium periodically. During callus formation there is some degree of dedifferentiation (*i.e.*, the changes that occur during development and specialisation are, to some extent, reversed), both in morphology (callus is usually composed of unspecialised parenchyma cells) and metabolism. One major consequence of this dedifferentiation is that most plant cultures lose the ability to photosynthesise.

This has important consequences for the culture of callus tissue, as the metabolic profile will probably not match that of the donor plant. This necessitates the addition of other components—such as vitamins and, most importantly, a carbon source—to the culture medium, in addition to the usual mineral nutrients. Callus culture is often performed in the dark (the lack of photosynthetic capability being no drawback) as light can encourage differentiation of the callus. During long-term culture, the culture may lose the requirement for auxin and/or cytokinin.

This process, known as 'habituation', is common in callus cultures from some plant species (such as sugar beet). Callus cultures are extremely important in plant biotechnology. Manipulation of the auxin to cytokinin ratio in the medium can lead to the development of shoots, roots or somatic embryos from which whole plants can subsequently be produced. Callus cultures can also be used to initiate cell suspensions, which are used in a variety of ways in plant transformation studies.

DEMONSTRATION OF TISSUE CULTURE FOR TEACHING

The starting point for all tissue cultures is plant tissue, called an explant. It can be initiated from any part of a plant - root, stem, petiole, leaf or flower - although the success of any one of these varies between species. It is essential that the surface of the explant is sterilised to remove all microbial contamination. Plant cell division is slow compared to the growth of bacteria and fungi, and even minor contaminants will easily over-grow the plant tissue culture. The explant is then incubated on a sterile nutrient medium to initiate the tissue culture. The composition of the growth medium is designed to both sustain the plant cells, encourage cell division, and control development of either an undifferentiated cell mass, or particular plant organs.

The concentration of the growth regulators in the medium, namely auxin and cytokinin, seems to be the critical factor for determining whether a tissue culture is initiated, and how it subsequently develops. The explant should initially form a callus, from which it is possible to generate multiple embryos and then shoots, forming the basis for plant regeneration and thus the technology of micropropagation. The first stage of tissue culture initiation is vital for information on what combination of media components will give a friable, fast-growing callus, or a green chlorophyllous callus, or embryo, root or shoot formation. There is at present no way to predict the exact growth medium, and growth protocol, to generate a particular type of callus. These characteristics have to be determined through a carefully designed and observed experiment for each new plant species, and frequently also for each new variety of the species which is taken into tissue culture. The basis of the experiment will be media and protocols that give the desired effect in other plant species, and experience.

The Demonstration

The strategy for designing a medium to initiate tissue culture, showing how growth regulators and other factors modulate development, can be demonstrated using the African Violet, a popular house plant. Leaf sections are the source of explants. This demonstration is regularly carried out by a student class, and gives reliable results. Sterile supplies are provided from central facilities, and provision of sterile working areas (for example, in laminar flow hoods) is an advantage, although cultures can be initiated in an open laboratory with careful aseptic technique. The standard precautions used during any laboratory work involving chemicals or microbes should be adopted. If you are in any doubt about safety hazards associated with this demonstration, you should consult your local safety adviser.

Step 1: Selection of the Leaves

Leaves are cut from healthy plants, leaving a short length of petiole attached. They should be selected to each yield several explants of leaf squares with approximately 1 cm sides. The youngest and oldest leaves should be avoided. Wash the dust off the leaves in a beaker of distilled water, holding the leaf stalk with forceps.

Step 2: Surface Sterilisation and Preparation of the Explants

This part of the procedure should be carried out in a sterile working area, or with meticulous aseptic technique. The leaf, with the petiole still attached, should be immersed in 70 per cent ethanol for 30 seconds, then transferred to a sterile petri dish. Sterile scissors and forceps are then used to cut squares from the leaf as explants, each with approximately 1 cm sides.

The explants are transferred into a 10 per cent hypochlorite bleach solution for 5 minutes, gently agitating once or twice during this time. They are then washed free of bleach by immersing in four successive beakers of sterile distilled water, leaving them for 2-3 minutes in each. Three explants are placed on each petri dish of growth medium, with the upper epidermis pressed gently against the surface of the agar to make good contact. The petri dishes are sealed with plastic film to prevent moisture loss, and incubated at 25°C in 16h light/ 8h dark.

Step 3: Assessment of Tissue Culture Development

The explants are incubated for 4 - 6 weeks, and inspected at weekly or fortnightly intervals. The growth of obvious bacterial or fungal colonies indicates contamination, and data from such cultures is obviously suspect. The development of dark brown tissue cultures can also be a consequence of contamination.The media used in the demonstration are designed to show the effects of auxin, cytokinin, sucrose and mineral salts on development. The media were based on the well-known Murashige and Skoog inorganic medium.

Typical Results

Absence of sucrose inhibits this production. Shoot production is also limited on the low sucrose concentration, but comparable with the control at high sucrose.At zero and low levels of cytokinin, callus forms where the leaf surface is in contact with the medium, while at high levels, shoot formation is stimulated.At zero and low levels of auxins there is a stimulus to shoot formation, but at high concentrations, large numbers of roots are formed.At low and zero levels of MS salts, there is no growth at all.These very obvious variations demonstrate the importance of a carbon and inorganic salt source for plant growth, as well as the effect of the auxin:cytokinin ration on the control of plant development.

REGULATORY MUTATIONS AND THERE BEHAVIOUR

If a stem is to branch, new shoot apical meristems must be created, and this too depends on events in the neighbourhood of the shoot apex. At each developing node, in the acute angle (the axil) between the leaf primordium and the stem, a bud is formed. This contains a nest of cells, derived from the apical meristem, that keep a meristematic character.

They have the capacity to become the apical meristem of a new branch or the primordium of a structure such as a flower; but they also have the alternative option of remaining quiescent as *axillary buds*. The plant's pattern of branching is regulated through this choice of fate, and mutations that affect it can transform the structure of the plant. Maize provides a beautiful example.

Maize represents one of mankind's most remarkable feats of genetic engineering. Native Americans created it by selective breeding, over a period of several centuries or perhaps millennia between 5,000 and 10,000 years ago. They started from a wild grass known as teosinte, with highly branched leafy stems and tiny ears bearing inedible hard kernels.

Detailed genetic analysis has identified a handful of genetic loci—about five—as the sites of the mutations that account for most of the difference between this unpromising ancestor and modern corn. One of these loci, with a particularly dramatic effect, corresponds to a gene called *teosinte branched-1* (*tb1*). In maize with loss-of-function mutations in *tb1*, the usual simple unbranched stem, with a few large leaves at intervals along it, is transformed into a dense, branching, leafy mass reminiscent of teosinte. The pattern of branching in the mutant implies that axillary buds, originating in normal positions, have escaped from an inhibition that prevents them, in normal maize, from growing into branches.

Fig. Transformation of Plant Architecture. (A) Photographs of three types of plants. (B) The Architecture of Teosinte, Normal Maize and the *tb1*-Defective Maize Compared Schematically.

In normal maize, the single stem is crowned with a tassel—a male flower—while a few of the axillary buds along the stem develop into female flowers and, upon fertilization, form the ears of corn that we eat. In the mutant maize with a defective *tb1* gene, these fruitful axillary buds are transformed into branches bearing tassels. The wild teosinte plant is like the *tb1*-defective maize in its leafy, highly branched appearance, but unlike this mutant it makes ears on many of its side branches, as though *tb1* were active.

DNA analysis reveals the explanation. Both teosinte and normal maize possess a functional *tb1* gene, with an almost identical coding sequence, but in maize the regulatory region has undergone a mutation that boosts the level of gene expression. Thus in normal maize the gene is expressed at a high level in every axillary bud, inhibiting branch formation, while in teosinte the expression in many axillary buds is low, so that branches are permitted to form. This example shows how simple mutations, by switching the behaviour of meri-stem cells, can transform plant structure—a principle of enormous importance in the breeding of plants for food. More generally, the case of *tb1* illustrates how new body plans, whether of plant or animal, can evolve through changes in regulatory DNA without change in the characters of the proteins made.

HORMONAL SIGNALS AND COORDINATE DEVELOPMENTAL

The fate of an axillary bud is dictated not only by its genes, but also by environmental conditions. Separate parts of a plant experience different environments and react to them individually by changes in their mode of development. The plant, however, must continue to function as a whole.

This demands that developmental choices and events in one part of the plant affect developmental choices elsewhere. There must be long-range signals to bring about such coordination. As gardeners know, for example, by pinching off the tip of a branch one can stimulate side growth: removal of the apical meristem relieves the quiescent axillary meristems of an inhibition and allows them to form new twigs. In this case the long-range signal from the apical meristem, or at least a key component has been identified.

It is an auxin, a member of one of six known classes of plant growth regulators (sometimes called *plant hormones*), all of which have powerful influences on plant development. The five other known classes are the *gibberellins*, the *cytokinins, abscisic acid*, the gas *ethylene*, and the *brassinosteroids*.

As shown in Figure, all are small molecules that readily penetrate cell walls. They are all synthesised by most plant cells and can either act locally or be transported to influence target cells at a distance. Auxin, for example, is transported from cell to cell at a rate of about 1 cm per hour from the tip of a shoot towards its base. Each growth regulator has multiple effects, and these are modulated by the other growth regulators, as well as by environmental cues and nutritional status. Thus auxin alone can promote root formation, but in conjunction with gibberellin it can promote stem elongation, with cytokinin, auxin it can suppress lateral shoot outgrowth, and with ethylene it can stimulate lateral root growth.

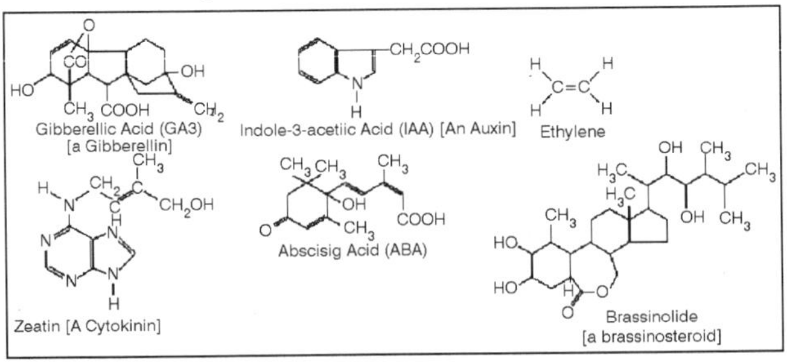

Fig. Plant Growth Regulators. The Formula of one Naturally Occurring Representative Molecule from Each of the Six Groups of Plant Growth Regulatory Molecules is Shown.

Homeotic Selector Genes Specify the Parts of a Flower

Meristems face other developmental choices besides that between quiescence and growth, as we have already seen in our discussion of maize, and these also are frequently regulated by the environment. The most important is the decision to form a flower.

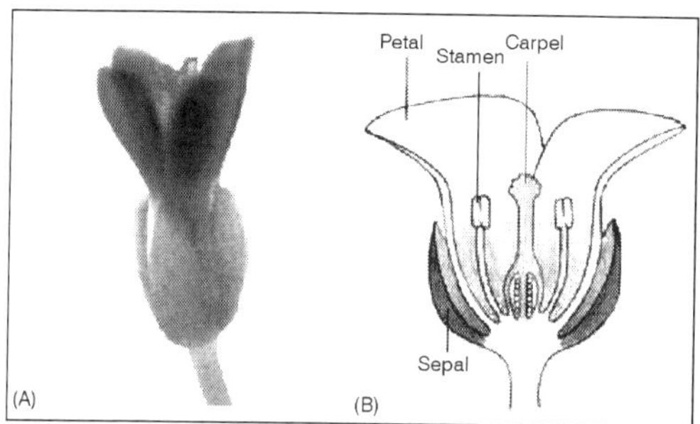

Fig. The Structure of an *Arabidopsis* flower. (A) Photograph. (B) Drawings. (C) Schematic Cross-sectional View. The Basic Plan, as Shown in (C), is Common to most Flowering Dicotyledonous Plants.

The switch from meristematic growth to flower formation is typically triggered by light. By poorly understood mechanisms based on light absorption by phytochrome and cryptochrome proteins, the plant can sense very precisely a change in day length. It responds by turning on expression of a set of *floral meristem-identity* genes in the apical meristem. By switching on these genes, the apical meristem abandons its chances of continuing vegetative growth and gambles its future on the production of gametes.

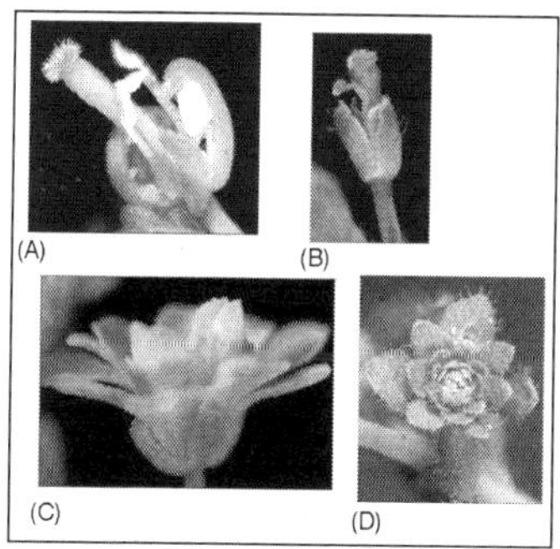

Fig. Arabidopsis Flowers Showing a Selection of Homeotic Mutations.

(A) In Apetala2, Sepals are Converted into Carpels and Petals into Stamens; (B) In Apetala3, Petals are converted into Sepals and Stamens into Carpels; (C)In Agamous, Stamens are Converted into Petals and Carpels into Floral Meristem. (D) In a Triple Mutant.

Its cells embark on a strictly finite programme of growth and differentiation: by a modification of the ordinary mechanisms for generating leaves, a series of whorls of specialised appendages are formed in a precise order—typically sepals first, then petals, then stamens carrying anthers containing pollen, and lastly carpels containing eggs. By the end of this process the meristem has disappeared, but among its progeny it has created germ cells. The series of modified leaves forming a flower can be compared to the series of body segments forming a fly. In plants, as in flies, one can find homeotic mutations that convert one part of the pattern to the character of another. The mutant phenotypes can be grouped into at least four classes, in which different but overlapping sets of organs are altered).

The first or 'A' class, exemplified by the *apetala2* mutant of *Arabidopsis*, has its two outermost whorls transformed: the sepals are converted into carpels and the petals into stamens. The second or 'B' class, exemplified by *apetala3*, has its two middle whorls transformed: the petals are converted into sepals and the stamens into carpels. The third or 'C' class, exemplified by *agamous*, has its two innermost whorls transformed, with a more drastic consequence: the stamens are converted into petals, the carpels are missing, and in their place the central cells of the flower behave as a floral meristem, which begins the developmental performance all over again, generating another abnormal set of sepals and petals nested inside the first and, potentially, another nested inside that, and so on, indefinitely. A fourth class, the *sepallata* mutants, has its three inner whorls all transformed into sepals. These phenotypes identify four classes of homeotic selector genes, which, like the homeotic selector genes of *Drosophila*, all code for gene regulatory proteins.

These are expressed in different domains and define the differences of cell state that give the different parts of a normal flower their different characters, as shown in Figure. The gene products collaborate to form protein complexes that drive expression of the appropriate downstream genes. In a triple mutant where the A, B and C genetic functions are all absent, one obtains in place of a flower an indefinite succession of tightly nested leaves. Conversely, in a transgenic plant where genes of the A, B and *sepallata* classes are all expressed together outside their normal domains, leaves are transformed into petals. Leaves therefore represent a "ground state" in which none of these homeotic selector genes are expressed, while the other types of organ result from expressing the genes in different combinations.

BASIC STEPS OF PLANT TISSUE CULTURE

Some basic steps are followed in plant tissue culture which is described below.

Explant Selection

Explant is the plant fraction used for tissue culture. Explant from healthy and young part of the plant is used. Parenchymas from stems, rhizomes, tubers, roots are easily accessible and respond quickly to culture condition.

Sterilisation Elimination

Sterilisation means elimination of all living organisms (microbes). The tissue culture is carried out in completely aseptic condition for which there is a need to properly sterilise all the glasswares culture media and explants. The explants are surface sterilised by repeated washing in sterile water and by using disinfectants such as mercuric chloride, hydrogen peroxide, etc. The glassware, culture media and other instruments are sterilised in an instrument called autoclave where sterilisation is done in steam under high pressure or in a pressure cooker.

Preparation of Nutrient or Culture Medium

The culture medium is prepared in aseptic condition. The basic constituents of any culture medium are:

- *Inorganic nutrients:* Inorganic nutrients include macronutrients as salts of nitrogen, phosphorous, potassium, calcium, magnesium and sulphur and micronutrients like boron, molybdenum, copper, zinc. Iron and chloride

Source of Carbon

Sucrose is used as the source of carbon. Several growth hormones like 2,4,1), cytokinins-benzylaminopurine, kinetin, myoinositol, 1AA, XAA arc used, vitamins like nicosstinic acid, pyridoxinc-IICL are added to the medium. After the addition of all the ingredients in appropriate proportion, agar-agar is added to prepare a solid medium. In some types of culture (like root culture) liquid medium is used (no agar-agar used).

Process of Inoculation

Inoculation is the process of the transfer of explant to suitable nutrient medium contained in culture vessels. This is done in sterile condition either in an inoculation chamber or laminar flow. After the inoculation the culture vessels are maintained in controlled temperature and light. The suitable temperature, for tissue culture ranges between 18-25°C.

Callus Configuration and its Culture

A callus is an amorphous mass of loosely arranged thin walled parenchyma cells developing from proliferating cells of parent tissue (Dodos and Roberts, 1985). The nutrient medium supplemented with auxins induces cell division and soon the upper surface of explant is covered by callus. The callus has the biological potential to develop normal root, shoots and ultimately forming a plant. Callus is formed through three developmental stages: induction, cell division and differentiation. Callus formation is governed by the source of explant, nutritional composition of the medium and environmental factors.

During the induction period the metabolic rate of cells is stimulated. Owing to increased metabolic rate cells enter to cell division stage. In the third phase, cellular differentiation and expression of certain metabolic pathways start leading to secondary products. When callus are grown on a nutrient medium for a long time, it becomes essential to subculture it in fresh media.

Development of Organogenesis

Organogenesis means the development of organs like root, shoot and leaves (but not embryo). Organogenesis starts with stimulation caused by the chemical of medium, substances carried over from the original explants and endogenous compounds produced by the culture. Organogenesis can be induced with the application of varying proportions of auxin and cytokinin. Skoog and Miller (1957) demonstrated that a high ratio of auxin: cytokinin (3: 0.02) stimulated root formation in tobacco callus but a low ratio of the same (3: 0,2) induced shoot formation.

Formation of Somatic Embryogenesis

This is the process of inducing embryo formation from somatic cells of cultured plant tissue. The embryo thus developed is knowTi as embryoids. Two different nutritional media are required to obtain embryoids. First medium contains auxin to initiate embryogenic cells. Second medium lacks auxin or has reduced level of auxins for subsequent development of embryonic cells into embryoids and plantlets.

The embryogenic cells pass through three different stages, *e.g.*, globular, heart shaped and torpedo shaped to form embryoids. Some plants in which somatic embryogenesis has been induced *in vitro* are Atropa belladona, Brassica oleracea, Carica papaya, Coffea arabica, Citrus cinensis, Daucus carroia, Nicotiana tobacum, etc.

Somaclonal Variations: In 1981, P.J Larkins and W. R. Scowcrott at the Division of Plant Industry, C.S.I.R.O. Australia gave the name somaclonal variation to genetic variability generated during tissue culture.

Though the cultured tissues are grown from single explant, over a long period of maintenance genetic variabilities are marked in the cultures.

This may be due to:

- Reflection of heterogeneity between cells and explant tissues,
- A simple representation of spontaneous mutation rate or
- Activation by culture environment of transposition of genetic materials.

Somaclonal variants of leaf callus culture of Solanum tuberosum have shown characters like disease resistance, variations in maturity dates of tubers, yield and shape.

Cell suspension cultures: Cell suspension is prepared by transferring a fragment of callus (about 500 mg) to liquid medium (500 ml) and agitating them aseptically in a shaker to make the cells free. The suspension then includes single cell, cell aggregates, and residual including dead cells. A good suspension contains high proportion of single cells.

Cell suspension cultures have many advantages over the callus cultures as below:

- The cell suspension can be pipetted.
- They are less heterogeneous.
- They can be cultured in volumes up to 1,500 litres.
- They can be subjected to more stringent environmental controls.
- They can be manipulated for production of natural products.

THEORY OF TISSUE CULTURE

The theoretical basis for plant tissue culture was proposed by Gottlieb Haberlandt, German Academy of science in 1902 on his experiments on the culture of single cell. The first true cultures were obtained by Gautheret from cambial tissue of Acer pseudoplatanus. The term plant tissue culture (Micro propagation) is generally used for the aseptic culture of cells, tissues, organs and their components under defined chemical and physical conditions *in vitro*. The basic concept of the plant body can be dissected into smaller part termed as "explants" and any explants can be developed into a whole plant. It is a central innovative areas of applied plant science, including agriculture and plant biotechnology. This technique is effective because almost all the plants cell are totipotent; In each cell possesses the genetic information and cellular machinery necessary to generate the whole organism. Since, this technique can be used to produce a higher number of plants that are genetically similar to a parent plant as well as to another. Two concepts, plasticity and totipotency, are the central processes to understand the regeneration and plant cell culture. Plants, due to its longer life span and sessile nature, have developed a greater ability to overcome the extreme conditions.

Most of the processes include in plant development and the growth, adapt to environmental conditions. When the plant cells and tissues are cultured *in vitro*, most of them are generally exhibit a very high degree of plasticity, which allows one type of organ or tissue to be initiated from another type. Like this way, the whole plant can be subsequently regenerated. These maintenance of genetic potential is called totipotency. The plant tissue culture medium is an artificial nutrient supplement of organic and inorganic nutrients used for cultivation of plant tissue media. The appropriate composition of the medium largely determines the success of the culture. The culture media used for the *in vitro* cultivation of the plant cells are composed of three basic components.

- Essential elements (normal ions) supplied as a complex mixture of salts.
- An organic supplements providing vitamins and amino acids.
- A source of fixed carbon which is usually supplied as sucrose.

When cultured in an appropriate medium having auxin and cytokinin, explants will give rise to an unorganised, growing and dividing mass of cells called callus. Callus cultures are initiated from a small part of an organ or tissue segment called the explants on a growth supporting solidified nutrient medium under sterile conditions. Any part of the plant organ or tissues may be used as the explants. At the time of callus formation, there is some degree of dedifferentiation happens both in morphology and metabolism. One of the major consequences of this dedifferentiation is that most plant cultures lose their ability to perform photosynthesis. The necessitates of the addition of other components such as carbon and vitamins source to the culture media, in addition to the unusual mineral nutrients.

MORPHOLOGY OF CALLUS

Callus varies considerably in appearance and texture, ranging from hard nodular cell masses to friable soft ones. They maybe white or creamish, orange, green either in whole or part as a result of chloroplast development. The shape of individual cells within the callus mass ranges from the near spherical or markedly elongated. A typical unorganised plant callus initiated from a new explants or piece of previously initiated calli has three stages of development.

1. The induction of cell division.
2. A period of active cell division during which differentiated cells lose specialised features they may have acquired and become dedifferentiated. Cell division usually occurs in the outer layer of the explants.
3. Period when cell division slows down on ceases and when within the callus, there is increasing cellular differentiate.

Callus culturing is performed in the dark while light can be encourage the differentiation of the callus. At the time of long term culture, the culture may loss the requirement for cytokinin and auxins. Manipulation of the auxins to cytokinin ratio in the medium can leads to the development of shoots, roots or somatic embryos from which the plant can be subsequently produced.

Callus culture is useful for many purposes:

- Callus is the starting material for the suspension culture which cells are separated.
- It helps in the production of secondary plant products.
- It is useful for the synthesis of starting compounds that are subsequently modified to yield the desired product.
- It is the starting materials for vegetative propagation of plants.

Based on the availability of the various *invitro* techniques, the dramatic increase in their application to various problems in basic biology, agriculture, horticulture, and forestry.

The applications can divide conveniently into six broad areas:

1. Cell behaviour.
2. Plant modification.

3. Germplasm storage and pathogen –free plants.
4. Clonal propagation.
5. Product formation.
6. Improved varieties.

PLANT TISSUE CULTURE

Plant cells can be grown in isolation from intact plants in tissue culture systems. The cells have the characteristics of callus cells, rather than other plant cell types. These are the cells that appear on cut surfaces when a plant is wounded and which gradually cover and seal the damaged area.

Pieces of plant tissue will slowly divide and grow into a colourless mass of cells if they are kept in special conditions. These are:

- initiated from the most appropriate plant tissue for the particular plant variety
- presence of a high concentration of auxin and cytokinin growth regulators in the growth media
- a growth medium containing organic and inorganic compounds to sustain the cells
- aseptic conditions during culture to exclude competition from microorganisms

The plant cells can grow on a solid surface as friable, pale-brown lumps (called callus), or as individual or small clusters of cells in a liquid medium called a suspension culture. These cells can be maintained indefinitely provided they are sub-cultured regularly into fresh growth medium. Tissue culture cells generally lack the distinctive features of most plant cells.

They have a small vacuole, lack chloroplasts and photosynthetic pathways and the structural or chemical features that distinguish so many cell types within the intact plant are absent. They are most similar to the undifferentiated cells found in meristematic regions which become fated to develop into each cell type as the plant grows. Tissue cultured cells can also be induced to re-differentiate into whole plants by alterations to the growth media.

Plant tissue cultures can be initiated from almost any part of a plant. The physiological state of the plant does have an influence on its response to attempts to initiate tissue culture. The parent plant must be healthy and free from obvious signs of disease or decay. The source, termed explant, may be dictated by the reason for carrying out the tissue culture. Younger tissue contains a higher proportion of actively dividing cells and is more responsive to a callus initiation programme. The plants themselves must be actively growing, and not about to enter a period of dormancy. The exact conditions required to initiate and sustain plant cells in culture, or to regenerate intact plants from cultured cells, are different for each plant species. Each variety of a species will often have a particular set of cultural requirements. Despite all the knowledge that has been obtained about plant tissue culture during the twentieth century, these conditions have to be identified for each variety through experimentation.

HISTORY OF PLANT TISSUE CULTURE

Plant tissue culture, an off-shoot of human curiosity has presently become an essential component of plant biotechnology. Through the efforts of pioneers in France, USA, U.K. and Germany, plant tissue culture Trichopus zeylanicus (locally called arogya-pacha) is a wild plant that occurs in the Agasthyamalai hills of Kerala. It yields a drug that acts as an anti-fatigue and restorative agent. Scientists of Regional Research Laboratory (RRL), Jammu, and later of Tropical Botanical Garden and Research Institute (TBGRI), Kerala have carried our ethno-pharmacological studies on this plant and have prepared a formulation called Jeewani.

The technology has been transferred to Arya Vaidya Pharmacy, Coimbatore. Fifty per cent of the monetary benefit from this transaction is being shared with the Kani tribe. This is the first example in which ethnobiological studies have resulted in tangible benefits to a tribal community. There are a large number of wild plants with potential use in agriculture, horticulture and medicine, requiring attention. spread to various parts of the world. It was P. Maheshwari who started the first tissue culture Laboratory in India at DU in the 1950s as he foresaw the value of this technique in experimental embryology. Teachers trained abroad returned to Delhi, Baroda and Pune from where the interest spread to other universitites and CSIR laboratories such as NCL, NBRI, RRL(Jammu), and to BARC, Mumbai. The main benefits derived from tissue culture are: control of organogenesis, elimination of breeding barriers, micropropagation, disease detection and eradication, somatic embryogenesis, use of protoplasts, somatic hybridization, somaclonal variation, detection of genetic variability in vitro and production of transgenics.

The discovery of androgenic haploidy at DU opened up an entirely new field of research in the production of pure lines of crop plants resulting in reduction of time and labour for exploiting hybrid vigour, especially for rice in China and later for other cereals. Very recently gynogenic haploids have been produced in mulberry. Undeniably the most useful outcome of tissue culture has been in the micropropagation of ornamentals, agricultural and plantation crops, fruit and forest trees. Laboratory research on micropropagation of plants of economic importance (Citrus, Eucalyptus, bamboos, teak, poplar, banana, sugarcane, turmeric and cardamom) has been scaled up to near commercial level. The Asian bamboos have long flowering cycles and come to mast seeding once in 12-120 years. The first demonstration of micropropagation in bamboo using seed callus cultures was done in Delhi and over 10,000 plants were transferred to the field successfully. Several groups working in India and abroad have now demonstrated that micropropagation in bamboos can be accomplished, starting from vegetative parts. The pioneering work done at Delhi and NCL, Pune, led to the discovery that bamboo plantlets can come to flower in vitro precociously. Tissue culture has been effectively used for multiplying and storing economically important, endangered (*e.g.* Nepenthes khasiana), threatened (Himalayan orchids) and biologically incompletely understood plants (Podostemaceae) and medicinal plants (Dioscorea, Coptis teeta, Valeriana wallichi, Podophyllum hexandrum, Picrorhiza kurroa) at North-Eastern Hill University (NEHU), DU, NBRI, NCL, BARC, CIMAP and Institute of Himalayan Bioresources and Technology (IHBT), Palampur.

The DBT has established six centres in India to provide hardening facilities for laboratory-raised plants and Micropropagation Technology Parks at TERI and NCL. The National Facility for Plant Tissue Culture Repository (also started by DBT) located at the NBPGR, has now been taken over by the ICAR. Collection, evaluation, introduction, exchange and conservation of germplasm, storage of elite plants, rare hybrids, and germplasm of vegetatively propagated plants and of plants bearing recalcitrant seeds in the form of tissues or embryonal axes in cryobanks are being practised at NBPGR.

TECHNIQUES OF PLANT TISSUE CULTURE

Modern plant tissue culture is performed under aseptic conditions under HEPA filtered air provided by a laminar flow cabinet. Living plant materials from the environment are naturally contaminated on their surfaces (and sometimes interiors) with microorganisms, so surface sterilization of starting material (explants) in chemical solutions (usually sodium or calcium hypochlorite or mercuric chloride) is required.

Mercuric chloride is seldom used as a plant sterilant today, unless other sterilizing agents are found to be ineffective, as it is dangerous to use, and is difficult to dispose of. Explants are then usually placed on the surface of a solid culture medium, but are sometimes placed directly into a liquid medium, particularly when cell suspension cultures are desired. Solid and liquid media are generally composed of inorganic salts plus a few organic nutrients, vitamins and plant hormones. Solid media are prepared from liquid media with the addition of a gelling agent, usually purified agar. The composition of the medium, particularly the plant hormones and the nitrogen source (nitrate versus ammonium salts or amino acids) have profound effects on the morphology of the tissues that grow from the initial explant.

For example, an excess of auxin will often result in a proliferation of roots, while an excess of cytokinin may yield shoots. A balance of both auxin and cytokinin will often produce an unorganised growth of cells, or callus, but the morphology of the outgrowth will depend on the plant species as well as the medium composition. As cultures grow, pieces are typically sliced off and transferred to new media (subcultured) to allow for growth or to alter the morphology of the culture. The skill and experience of the tissue culturist are important in judging which pieces to culture and which to discard. As shoots emerge from a culture, they may be sliced off and rooted with auxin to produce plantlets which, when mature, can be transferred to potting soil for further growth in the greenhouse as normal plants.

PLANT TISSUE CULTURE REQUIREMENT

In order to initiate plant tissue culture, one needs; explants, suitable culture/growth media, aseptic conditions to curb growth of microorganisms, water, growth regulators (auxins and cytokinins), and frequent sub culturing to avoid accumulation of waste metabolite and enhance nutrition.

Culture Medium

Plant tissue culture medium that is meant for cultivation of plant cells *in vitro* should contain mineral ions or essential elements in form of a mixture of complex salts, a source of carbon which is usually sucrose and organic supplements that supply vitamins and amino acids.

Organic Supplements

Organic supplements supply vitamins and amino acids. The two main vitamins essential for *in vitro* tissue culture are myo-inositol and thiamine while the most important amino acid is glycine.

Carbon Source

Sucrose is the preferred carbon source essential in a culture medium. It is easily available, cheap, easily assimilated and stable.

Essential Elements/Nutrients

Essential elements are classified as macronutrients, micronutrients and an iron source. A combination of these elements is necessary for tissue culture. Macronutrients are those elements that are supplied in large amounts for plant growth and development. They include; magnesium, sulphur, calcium, nitrogen, potassium, phosphorus and carbon which supplied separately.

All these elements comprise more than 0.1 per cent of plants' dry weight. Nitrogen is mainly supplied as nitrate ions or ammonium ions. However, high concentrations of ammonium ions lead to acidification of the medium and increase vitrification. Microelements are those elements that are needed in trace amounts in tissue culture media. However, they have diverse functions in plant growth and development. These elements include; iron, molybdenum, Manganese, iodine, cobalt, copper, boron and zinc.

ROLE OF GROWTH REGULATORS IN TISSUE CULTURE

Due to the totipotency and plasticity of plant cells, certain manipulations to culture media are essential to determine certain developmental pathways of a plant cell. Plant hormones and their respective synthetic analogues are used as plant growth regulators. These growth regulators include;

Auxins-promote cell division and growth in explants. They support callus induction hence growth. However, high levels of auxins suppress organised growth promoting growth of meristem-like cells. The mostly used type of auxin for tissue culture is called 2, 4-Dichlorophenoxyacetic acid (2, 4-D).

Cytokinins-these are purine derivatives that support cell division. The two main types of cytokinins used in tissue culture are benzylaminopurine (BAP) and kinetine.

Gibberellins-these are naturally occurring compounds that are used in regulating plant cell elongation. GA3 is the most commonly used type of gibberellin in this technique.

Abscisic acid (ABA)-inhibits cell division in plants. It is mostly used in somatic embryogenesis to promote specific developmental pathways.

Auxins and cytokinins, as growth regulators, have basic roles to play in plant tissue culture. Often, they are used together but with different concentration rations which subsequently determine the type of culture regenerated. A high cytokinin to auxin ratio supports formation of shoots, whereas a high auxin to cytokinin ratio favours formation of roots. A balanced ratio favours production of callus.

Micropropagation procedure:

- Selection of an explant from a 'mother plant' that is healthy and vigorous. Usually, apical buds are preferred as explants but any other tissue can be used.
- Establishment of this explant in a plant culture medium. A medium supports growth and cell division. Depending on the plant requirement, different types of media are used for specific types of plants.
- Multiplication. In this stage, the explants give rise to a callus.
- Differentiation and
- Organogenesis

Culture Types

Cultures are produced from 'explants'. There are different types of cultures produced from explants depending on the conditions availed. They include;

Callus

This is an unorganised, growing and actively dividing mass of cells produced when both auxins and cytokinins are present in a culture medium, a procedure carried out in the dark to discourage differentiation. During formation of callus, there is morphological and metabolic dedifferentiation. Dedifferentiation results into inability of these cultures to photosynthesize hence attain a different metabolic profile from the 'mother plant'. This feature precipitates addition of other culture components. Manipulation of auxin to cytokinin ratios dictates root, shoot and somatic embryo development from which plants are produced. Callus cultures are classified as either compact or friable. Callus formation plays a central role in plant biotechnology.

Cell- suspension Cultures

These cultures are produced from friable callus placed in a liquid medium and agitated. This releases single cells into the medium which under correct conditions, grow and divide to produce cell-suspension cultures. These cells are maintained as batch cultures in flasks.

Protoplasts

These are plant cells without cell walls. Removal of cell walls can be done either mechanically or by use of enzymes. The former method results in poor quality yields while the latter yields high and pure cells. The liquid medium used is not agitated to avoid damaging the protoplasts. However, the medium is put maintained under high osmotic pressure and shallow to allow aeration. Organogenesis or somatic embryogenesis can be used to produce whole plants on solid media. Many transformations are done through this method.

Embryo Culture

Embryos are used to produce either a callus culture or a somatic embryo. An immature embryo from an embryogenic callus is the most recommended for regeneration of monocot plants.

Other culture types include; microspore culture, root cultures and shoot tip and meristem cultures. These cultures give rise to plant regeneration.

2

Plant Improvement through Tissue Culture

In introducing this research update, it was mentioned that the major impact of tissue culture technology would not be in the area of micropropagation, but rather in the area of controlled manipulations of plant germplasm at the cellular level. The ability to unorganise, rearrange, and reorganise the constituents of higher plants has been demonstrated with a few model systems to date, but such basic research is already being conducted on ornamental trees and shrubs with the intent of obtaining new and better landscape plants.

PLANT IMPROVEMENT IN CELL WALLS

Protoplasts are naked cells that lack cell walls. They are spherical with a plasmolysed cell content and are contained within a plasmalemma. In principle, each individual protoplast can reform a cell wall, and later initiate either a callus through sustained divisions, or an embryo, defined as a somatic embryo. In banana they are obtained from *in vivo* tissues or *in vitro* cultures.

This phenomenon is termed totipotency and denotes the recovery of a whole organism from a single cell, and is also applicable to protoplasts. In theory all cells are totipotent, but in practice it depends on the past cellular environment. Usually the morphogenetic competence is retained at the unicellular stage that corresponds to a protoplast. However since protoplasts are not exposed to the stabilising and inductive influence from neighbouring cells, they may have lost their plant regeneration capacity. In fact, factors such as the genotype or species of the manipulated plant, and the ontogenetic state of the explant source, exert a powerful effect on the regeneration potential of protoplasts.

Consequently the development of appropriate *in vitro* conditions for protoplast regeneration is complicated. Notwithstanding this, successful efforts have been made in isolation, cultivation and regeneration of protoplasts, since Nickell and Torrey pointed out their merit for crop improvement. Various tissues and an increasing number of plant species and genotypes have been successfully used in protoplast culture, but so far sufficient quantities of protoplasts for practical applications are not routinely met. While the value for agriculture of protoplasts still needs to be demonstrated, they are an invaluable tool for studies on permeability of ions and solutes, photosynthesis, phytohormones, phytochrome, and maintenance of totipotency. Moreover, protoplasts are useful for the uptake of foreign genetic material and to produce somatic hybrids through protoplast fusion. In addition, protoplasts are an excellent system for studies on cell genetics and even for plant virology.

These studies rely on isolated, clean and healthy protoplasts, which requires the appropriate choice of osmoticum, hydrolysing enzymatic solution, and pretreatment of the donor explant either *in vivo* or *in vitro*. This involves the right choice of explant age, cold treatment, phytohormone pretreatment, light intensity or photoperiod, and subculture rhythm, which affects both the internal metabolic status of the cells and cell-wall composition. Prior to cultivation isolated protoplasts need to be freed from enzymatic remains and debris, which are considered to be toxic. Culture in or on solid medium is considered to be more advantageous than in liquid medium because the development of a single protoplast into a colony can be followed up much better. For development, different factors such as medium composition are crucial, especially the nature and quantity of growth hormones.

Other important factors are the physical environment, plating density, and embedding conditions or the use of feeder layers, which all contribute in one way or another to cell-wall regeneration and cell division. Once cell divisions start, the level of auxin(s), as well as the colony density, usually need to be reduced to avoid overcrowding. This is done by subculturing, which gives the additional benefit that an adequate nutrient supply is maintained. Finally, for rooting, plants are usually transferred to cytokinin-free medium, possibly containing some auxin, and exposed to high light intensities, usually in illuminated plant growth chambers. During recent decades, plant biology workers have recognised the potential of protoplasts in many experimental systems, since an efficient enzymatic method for protoplast isolation was first established by Cocking.

SIGNIFICANCE AND USE OF PROTOPLASTS IN BANANA

Banana is now easily amenable to *in vitro* culture, and plants are regenerated from various explants through organogenesis, embryogenesis, anther culture and even from cultured protoplasts. This has created opportunities for other biotechnological applications. Cell suspensions, for example, opened the way for gene transfer and improved the production, the quality, and therefore the manipulation of cell suspension-derived protoplasts. Through genetic engineering it has become possible to confer new traits on banana plants, using either particle bombardment or Agrobacterium-mediated transfer. Protoplasts facilitate the direct transformation of plant cells by DNA microinjection and electroporation. However, many characters of agricultural interest are multigenic or ill-defined, and current transformation methodologies allow the integration of only a few foreign genes. Protoplast fusion, however, allows the transfer of several useful characters, even if detailed genetic or molecular knowledge of genes encoding for these desired characters is lacking. Protoplast fusion is therefore a complementary tool to increase nuclear and cytoplasmic variability and to confer desirable agronomic traits.

Important banana cultivars are susceptible to many pests and diseases, particularly Mycosphaerella fijiensis and Fusarium oxysporum f.sp.cubense, nematodes and insects. Very interesting traits of resistance have been identified in this genus for most serious diseases. However cross-breeding is very difficult in the genus since most edible cultivars are sterile even after profuse pollination by wild pollen-fertile genotypes. Protoplast fusion is therefore an option, as it can overcome such sexual barriers which occur frequently in banana, and which cannot be bypassed even through embryo rescue. Therefore somatic hybridisation between wild and cultivated banana is expected to produce hybrids combining agronomic traits with genetic resistance to pests and pathogens.

Protoplast fusion has the added benefit that it creates the possibility of generating asymmetric fusions, whereby selected cytoplasmic organelles, chromosomes or chromosome fragments from an irradiated protoplast donor could be combined with the genome of an acceptor protoplast. This strategy is of especial interest in banana, where the subspecies balbisiana is considered to be a source of multiple resistance and therefore could be partially utilised to improve banana cultivars, of which the most important component is the subspeciesacuminata. Protoplast fusion also makes the creation of synthetic triploid banana genotypes possible by, for example, combining haploid protoplasts derived from anther culture with protoplasts from a diploid improved cultivar.

Chimerism and somaclonal variation are factors that seriously limit rapid clonal propagation of banana. By their nature, protoplasts avoid chimerism because they originate from a single cell, and protoplasts may be use to dissociate chimeric plants. On the other hand the cellular heterogeneity of protoplast populations can be useful for isolating somaclonal variants with improved characters such as increased yield or pathogenic resistance, so that somaclonal variation may also be exploited to provide new sources of genetic variability.

Banana protoplast isolation and culturing is nowadays routine, although protoplast research on banana started 15 years later than on model plants. Now it is regrettably underused because of the strong emphasis on molecular and genomic studies in banana. Nevertheless, protoplasts have much to offer for non-conventional banana breeding, since they overcome sexual incompatibilities at interspecific and even at the intergeneric level, and allow the incorporation of multigenic traits such as yield, resistance to stress, pests and diseases. Finally protoplast techniques may improve banana when other classical methods have failed.

WORK ON ISOLATED CELLS AND PROTOPLASTS

Early Experiments

The first attempts at banana protoplast culture were made in 1984 by Bakry. Various explants (leaf and sheath from *in vitro* banana plants, callus from floral explants, and bract and tepals from *in vivo* banana plants) of diploid and triploid banana plants belonging to the AA, BB, AAA and AAB groups were tested. Enzyme solutions were composed of a modified CPW (Cell and protoplast washing solution) salt solution containing 0.7 M mannitol, and a mixture of various enzymes at 0.1 to 5 per cent (w/v) such as:

- *Pectinases*: Macerozyme;
- *Hemicellulases*: Hemicellulase
- *Cellulases*: cellulase R-10 'Onozuka', Cellulysin and Driselase.

Table. Cell and Protoplast Washing (CPW) Solution

Component	Concentration (mg/l)
KH_2PO_4	27.2
KNO_3	100
$CaCl_2.2H_2O$	150
$Mg\ SO_4.7H_2O$	250
KI	0.16
$Cu\ SO_4.5H_2O$	0.025

The enzyme solution was adjusted to pH 5.6 with KOH and filter-sterilised (0.22 mm). Small explants of about 1-5 mm^3 were incubated for 9 h with shaking (80 r.p.m.) at 27°C in the dark. The mixture was sieved through an 83 mm metallic sieve, followed by dilution with KMC salt solution. After centrifugation at 100 g for 5 min, the pellet was diluted in a KMC solution. Whatever the plant material, enzymatic solution and plant preconditioning (light/dark), only isolated plasmolysed cells were obtained.

Positive results were obtained when nodular calli were used as starting material. Floral explants cultivated on MS medium complemented with 500 mg/l casein hydrolysate, 2 mg/l IAA, and 2 mg/l BAP lead to calli covered with nodular bodies. These calli were embryogenic. With the enzyme mixture 2.5 per cent cellulase R10 'Onozuka', 0.2 per cent hemicellulase, 0.3 per cent pectolyase Y23 and 0.6 per cent macerozyme, Bakry produced the first living banana protoplasts. After calcofluor staining, protoplasts were found to be cellulose-free, perfectly round in shape, variable in size and having a dense cytoplasmic content. Although the yield was quite low the protocol was reproducible.

This pioneering work was confirmed by Cronauer and Krikorian in 1986 (using proliferating shoot tips of the cultivar 'Lacatan' [AAA group]) and by Da Silva Conceicao in 1989. Later on regular progress was made by several scientists. Initially banana calli from protoplasts were developed in 1992, followed by plant regeneration directly from protoplasts, protoplast transformation and somatic hybridisation.

Explants used as a Source of Protoplasts

In banana, *in vivo* explants are not a good source of protoplasts. Although Matsumoto *et al.* obtained protoplasts from *in vivo* bracts, in most cases initiation started from *in vitro* cultures. Leaf explants, slices of shoot tissue, roots or callus were the first choice as starting material for protoplast isolation because of their convenience. In fact, protoplasts can be obtained in banana from almost any tissue, including young leaves, sheaths, bracts, roots, and callus, but yields depend on the explant source.

Yields are quite low with most tissues and range from 0.1 to 2.8 × 106 protoplasts per gram, but the protoplasts are unable to divide. However, cell suspension-derived protoplasts gave high yields which ranged from 4 to 5 × 105 per gram with 'Long Tavoy'; from 1.1 to 3 × 105per gram with malaccensis; 2 to 20 × 106 per gram with 'Maça' to 6.6 × 107 per millilitre of packed cell volume (PCV) with 'Bluggoe'. Essential was the embryogenicity of the suspension. The first sustained cell divisions in protoplast culture leading to calluses were observed by Megia *et al.* in 1992 with M. acuminata cell suspension-derived protoplasts. Megia *et al.* and Panis *et al.* improved the protocol. This was confirmed by Matsumoto *et al.* and Assani *et al.* who produced plants from such 'embryogenic protoplasts'. Consequently embryogenic cell suspensions are the starting material of choice.

Banana Genotypes Tested

Bananas are either triploid, diploid or tetraploid in order of importance for food production, but protoplasts have been obtained so far from triploid, diploid and haploid genotypes. Below are listed the cultivars and wild types tested; the letter A denotes the contribution of the subspecies acuminata and B the contribution of the subspecies balbisiana.

Triploid Genotypes

- Musa cavendishii L.cv North banana (AAA);
- 'Lacatan' (AAA)
- 'Nanicao' (AAA)
- 'Bluggoe' (ABB)

- 'Maça' (AAB); 'Grande Naine' (AAA); 'Gros Michel' (AAA); 'Currare Enano' (AAB); 'Dominico' (AAB).

Diploid Genotypes

- 'Pisang Lilin' (AA)
- Musa acuminata ssp. burmanica type 'Long Tavoy' (AA)
- Musa acuminata ssp. malaccensis (AA)
- SF 265 (AA); IRFA 903 (AA); Col 49 (AA); 'Colatino Ouro' (AA); 'Pisang Klutuk Wulung' (BB); 'Pisang Klutuk' (BB); 'Pisang Batu' (BB) and 'Tani' (BB).

Protoplast Extraction and Purification

Enzymes

Various enzyme mixtures are used depending on the author and banana explant (CS = Cell Suspension; P = Plant explant; C = callus).

They are commonly composed of:

- Cellulases (Cellulase Onozuka R10, Cellulase Onozuka RS, Cellulysin, Macerase, Driselase)
- Pectinases (Pectinase; Macerozyme R10; Pectolyase Y 23)
- Hemicellulase (Rhozyme HP -150; Glusulase)

Protoplast Isolation Media

It is standard practice to incubate the plant material in the enzyme solution for a digestion period of 1-24 h either with shaking (40 r.p.m.) or no shaking, generally at 27°C in the dark. The best source of protoplasts is from embryogenic cell suspension cultivated in the light (30 or 65 mE m^{-2} s^{-1}) or darkness and regularly subcultured. Initiation should start 3-5 days, 7-10 days or 2-15 days after the last subculture, followed by sieving through a 200-400 mm mesh just before enzyme incubation.

Protoplast Purification

Protoplasts are purified from debris by sieving (100 mm and subsequently or directly through 56/32/25 mm sieves), which is preferred to flotation. With flotation purification, a 21 per cent sucrose solution and centrifugation at 120 g is used.

Protoplast Washing

After sieving, protoplasts are generally repeatedly washed to remove enzymes. Protoplast pellets obtained by centrifugation (50/66/90 g) are washed with protoplast isolation medium without enzymes; cell suspension culture medium without growth regulator plus 10 per cent mannitol; or 3 per cent $CaCl_2$ plus 0.5 per cent KCl solution; or 3 per cent $CaCl_2$ 0.5 per cent KCl solution and 10 per cent mannitol; or 0.6 M D-mannitol, 0.1 mM $CaCl_2$, 0.5 per cent PVP-40 and 3.5 mM MES.

Protoplast Quality Control

Complete cell wall degradation is confirmed both by the spherical shape of the released protoplasts and by the absence of fluorescence after staining with calcofluor white. Evans blue or FDA staining are commonly used for viability assessment.

Protoplast Culture Media

For banana protoplast culture, derived MS medium (supplemented with 2,4-D, or 2,4-D plus zeatin) or N-medium composed with N6 salt, vitamins, organic acid and sugar alcohol are commonly used. Due to difficulties in obtaining favourable protoplast development, many culture systems have been tested. Liquid and solid media (derived from MS) were initially tried for protoplast cultivation, but it was found that liquid cultures are not suitable for Musa protoplasts. In contrast, solid media supported sustained divisions. To overcome the low growth response, other media were tested, such as semi-solid media, protoplasts embedded in solid or semi-solid media, nurse cultures, and nurse cultures with a feeder layer.

Cocultivation in banana protoplast cultures was considered as it was reported that coculturing protoplasts of recalcitrant species, especially monocotyledons, with a reliable feeder culture, induced cell division. The banana protoplast literature describes various combinations of media and culture systems but, whatever the protocol, a high plating density (10^5-10^6/ml) was crucial, as well as the use of feeder layer, or nurse culture combined with a feeder layer. Protoplasts are cultured at 27 ± 1°C in the dark.

SELECTION OF PLANTS WITH ENHANCED STRESS OR PEST RESISTANCE

Perhaps the most heavily researched area of tissue culture today is the concept of selecting disease, insect, or stress resistant plants through tissue culture. Just as significant gains in the adaptability of many species have been obtained by selecting and propagating superior individuals, so the search for these superior individuals can be tremendously accelerated using *in vitro* systems. Such systems can attempt to exploit the natural variability known to occur in plants or variability can be induced by chemical or physical agents known to cause mutations.

All who are familiar with bud sports, variegated foliage and other types of chimeras have an appreciation for the natural variability in the genetic makeup or expression in plants. Chimeras are the altered cellular expressions which are visible, but for each of these which are observed many more differences probably exist but are masked by the overall organisation of the plant as a whole. For example, even in frost-tender species, certain cells or groups of cells may be frost hardy. However, because most of the organism is killed by frost, the tolerant cells eventually die because they are unable to support themselves without the remainder of the organised plant. Plant tissues grown *in vitro* can be released from the organisation of the whole plant through callus formation.

If these groups of cells are then subjected to a selection agent such as freesing, then those tolerant ones can survive while all those which are susceptible will be killed. This concept can be applied to many types of stress as well as resistance to fungal and bacterial pathogens and various types of phytotoxic chemical agents.

The goal of selecting such resistant cell lines would be to reorganise whole plants from them which would retain the selected resistance Current research in this area extends across many interests including attempts to select salt tolerant lines of tomato, freesing resistant tobacco plants, herbicide resistant agronomic crops, and various species of plants with enhanced pathogen resistance. Imagine, if you will, the impact of a fireblight-resistant Bartlett pear, a clone of pin oak for alkaline soils, or a selection of southern magnolia hardy to zone 4!

TISSUE CULTURE AND PATHOGEN FREE PLANTS

Another purpose for which plant tissue culture is uniquely suited is in the obtaining, maintaining, and mass propagating of specific pathogen-free plants. The concept behind indexing plants free of pests is closely allied to the concept of using tissue culture as a selection system. Plant tissues known to be free of the pathogen under consideration (viral, bacterial, or fungal) are physically selected as the explant for tissue culture. In most cases, the apical domes of rapidly elongating shoot tips are chosen.

These are allowed to enlarge and proliferate under the sterile conditions of *in vitro* culture with the resulting plantlets tested for presence of the pathogen (a procedure called indexing). Cultures which reveal the presence of the pathogen are destroyed, while those which are indexed free of pathogen are maintained as a stock of pathogen-free material. Procedures similar to these have been used successfully to obtain virus-free plants of a number of species and bacteria-free plants of species known to have certain leaf spot diseases. The impact of obtaining pathogen-free nursery stock can only be speculative, since little research documenting viral, bacterial, or fungal diseases transmitted through propagation of woody ornamentals is available.

SOMATIC HYBRIDISATION

The ability to fuse plant cells from species which may be incompatible as sexual crosses and the ability of plant cells to take up and incorporate foreign genetic codes extend the realm of plant modifications through tissue culture to the limits of the imagination. Most such manipulations are carried out using plant "protoplasts". Protoplasts are single cells which have been stripped of their cell walls by enzymatic treatment. A single leaf treated under these conditions may yield tens of millions of single cells, each theoretically capable of eventually producing a whole plant. This concept has fueled speculation as diverse as the possibilities of obtaining nitrogen-fixing corn plants on the one extreme to discovering a yellow-flowered African violet on the other extreme.

The observation that has provided the impetus for most of this research is that when cells are stripped of their cell walls and brought into close contact, they tend to fuse with each other. This "somatic hybridisation" is not subject to the same incompatibility problems that limit traditional plant breeding strategies. It is conceivable then that one could hybridise a Juneberry with a crabapple or a plum, but the fundamental research required to demonstrate such an event has yet to be conducted.

The potential use of somatic hybridisation to bring about novel combinations of genetic material has been demonstrated in the genera Petunia and Nicotiana. Research funded in part by the Horticultural Research Institute at the University of Wisconsin is investigating the feasibility of using such techniques with woody species. Brent McGown and co-workers have succeeded in obtaining naked cells from tissue cultures of Betula and Rhododendron, but as of yet, they have neither obtained plants from single cells not achieved cellular fusion.

However, further research in this area promises to have a tremendous impact on our concepts of woody plant diversity. Just as remarkable as the idea of fusing plant protoplasts is the idea of incorporating foreign genetic material into the genetic code of plant cells. Such transformations have been carried out in the so-called "gene-splicing" experiments where the information for making insulin was incorporated into bacteria. Not only is the desired information transmitted to succeeding generations of bacteria, but the bacterial cultures become synthesisers of insulin as well. Plant cells can be made to take up foreign genetic codes, but evidence that this can be transmitted into the daughter cells and serve the intended function is lacking. Plant tissue culture research is multi-dimensional. While most nurserymen have been introduced to the techniques and advantages of micropropagation, few have ventured to use it as a propagation tool.

IMPROVING TECHNIQUES OF THE QUALITY SYSTEM

The fertilized egg is bisected at the two celled stage and each half is transplanted into different regions of the uterus. Using this method the reproductive rate is doubled as we know that female sheep and cattle produce, on an average, one offspring per pregnancy. Hence the farmers can easily increase their farmstock. This technique can also be used to conserve rare breeds. After fertilizing their eggs in the laboratory, the young embryos or rare animals are dissected into anything from 2-8 cells. Each cell, is then transplanted into a surrogate mother of a common breed where it grows to produce a new individual of the rare type.

Another variation of this technique allows surrogate mothers to carry embryos of a different species, *e.g.* Horses giving birth to zebras from zebra embryos implanted at the blastocyst stage. The outer layer of zebra embryo is exchanged with the trophectoderm of the horse embryo in order to make it acceptable to the surrogate mother. This exchange of the layer does not affect the development of the inner cell mass which is the true embryo-forming region.

EMBRYO CLONING

Using this technique it is possible to produce many genetically identical copies of an animal. This method has been used to clone mouse however, research is going on to standardise this method to obtain clones of cattle with desirable traits, *e.g.*, cows with high milk production, etc.

Following steps are used in this method:

- An egg from the donor is grown to blastocyst stage under laboratory conditions.
- The egg is dissected to remove the inner cell mass.
- The mass cells are separated into individual cells.
- Injection of a nucleus from each of these cells into a one-celled embryo containing two pronuclei.
- Removal of pronuclei followed by culturing of embryos in the laboratory until the blastocyst stage.
- Transplantation of these embryos into surrogate mothers.

CHIMERAS

Early research into the production of transgenic animals revealed a simple method for producing hybrids, called Chimeras, between closely related species. Sheep-goat chimeras were produced by mixing four celled sheep embryos with eight-celled goat embryos. After removing the zona pellucida from each egg, the eggs were pressed together and incubated at 370C. The cells reorganised and formed hybrid blastocysts.

These hybrid blastocysts were then transferred to sheep foster mothers, where they continued their growth and development until the sheep gave birth. Each hybrid offspring contained both sheep and goat cells in all it's tissues which resulted in the coats of these animal having a patchwork kind of pattern with irregular patches of sheep and goat fur.

TISSUE CULTURE OF LILACS

Traditionally lilacs have been grown from seed, suckers or by grafting superior clones on to compatible rootstocks. Lilac seedlings of several species, privet stem cuttings as well as ash seedlings/root cuttings have all been utilized as rootstocks. In recent times there has been a move away from using rootstocks for various reasons, and nurserymen have been trying to produce cultivars on their own roots.

While cuttings taken at the right time are often successful for many species, this is not always a sure recipe for thevulgaris clones. For more than two decades some of the leading nurseries in the US and Europe have been using tissue culture for propagating lilacs on their own roots. While the main reason for resorting to micropropagation is to produce large numbers of selected clones within a short period, we have discovered it to be no less effective in producing small numbers ofvulgaris cultivars, which are recalcitrant to initiating roots by conventional means.

VITRO FLOWERING OF ROSE

The present invention is directed to compositions and methods of micropropagation of a rose plant and in vitro flowering of the rose plant. Young shoots are induced to produce buds in an enclosed vessel containing a first culturing medium comprising benzyladenine, an auxin and 2% sucrose as carbon source.

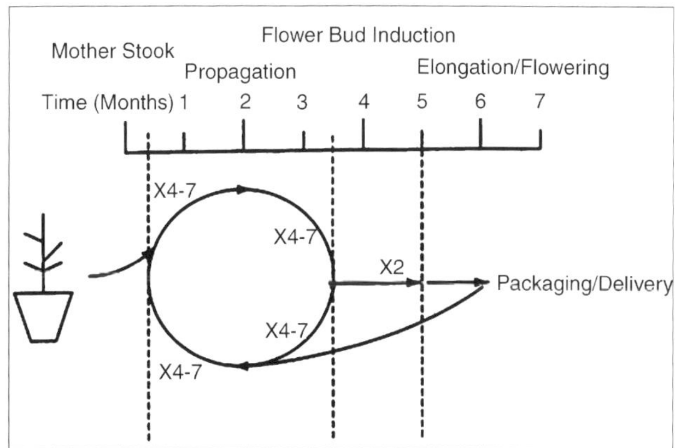

Buds are excised and cultured on a medium for propagation and multiplication of plantlets. Plantlets are then transferred to a medium comprising thidiazuron, an auxin and myo-inositol for induction of flower buds, followed by culture on a medium comprising benzyladenine and an auxin for plantlet elongation and finally culture in the fifth medium without phytohormone for flowering. Alternately, after propagation and multiplication, plantlets are transferred to a medium comprising zeatin, auxin and myo-inositol for induction of flower buds followed by culture on a medium without phytohormone for elongation and flowering.

ORGANOGENESIS

Meristematic tissues are responsible for the division of new cells... they are zones of actively dividing cells.

Before we proceed, we should first recognize that growth in plants includes two stages:

1. First the production of new cells and,
2. Secondly the expansion of these cells via uptake of water by the vacuole.

Cell division occurs solely in meristematic regions, while expansion may occur anywhere. Thus in a single plant there are zones of young dividing cells, maturing cells, and mature cells. We will go into more detail on meristematic regions when we go over the stem and root individually, but for now an introduction is sufficient

Let us recognize that there are 3 meristematic regions in the plant:

I. Apical meristems are located at the apices or tips - at root and shoot tips and are directly involved in their elongation

They create derivatives which form primary growth:

- The protoderm which forms the outer dermal layer of tissues,
- The ground meristem which forms the cortical cells and
- The procambium which forms the vascular tissue.

In shoots, plants protect their meristems with young leaves and by forming dormant reserve meristems (*i.e.*, buds), that protect their apical meristems with a root cap Ferns, Gymnosperms and Angiosperms are all terrestrial plants. Tissues therefore arise as a result of a cell or a group of cells dividing, thus giving rise to a large number of cells. In flowering plants these cells undergo changes that change the structure and function of the cell; differentiation has take place. As a result of differentiation, a cell or group of cells can have a specialized function; *i.e.* specialization takes place. Thus, a tissue is a group of cells that have differentiated to perform a specific function.

If a group of cells are more or the less the same, then it is known as a Simple tissue *e.g.* epidermis, parenchyma, collenchyma and sclerenchyma. If, however, the tissue is composed of a number of different cells, then it is known as a Complex tissue such as xylem and phloem.

DIAGRAMMATIC REPRESENTATION OF A PLANT BODY

Ground Tissue is simple non-meristematic tissue (non-dividing tissue) made up 3 cell types:paranenchyma, collenchyma and sclerenchyma.This tissue generally forms either the pith, cortex or bulk of leaf (mesophyll).

Parenchyma Cells

- Mostabundant cells in plants;
- Spherical cells which flatten at point of contact;
- Alive at maturity;pliable, primary cell walls;
- Large vacuoles for storage of starch, fats, and tannins (denature proteins);
- Primary sites of the metabolic functions such as photosynthesis, respiration, and protein synthesis; they are "ready reserves" from which a plant makes specialized cells to meet its changing needs.

SPECIALIZED PARENCHYMA

- *Chlorenchyma*: Photosynthetic cells; have high density of chloroplasts
- *Aerenchyma*: prominent intercellular spaces that improve gas exchange capacity of the tissue; provide maximum support with a minimum metabolic requirement
- *Transfer cells*: specialized for short distance transfer of solutes between cells; have secondary cell walls; they are inner extensions of wall that increase surface area.

Transfer cells occur in areas of high solute transport, such as secretory tissues, which release substances that are produced within the protoplasm and are moved outside, *i.e.*, nectar cells, mucilage in sundews, and resins.

Collenchyma Cells- Colla = Glue, Glisten

- Living protoplasm;
- Unevenly thickened primary cell walls; elongate cells;
- Longer than wide; just beneath epidermis; function to support growing organs, grass, floral parts, and border veins; their nonlignified cell walls can stretch

Sclerenchyma Cells- (skleos = hard)

- Most are dead at maturity;
- Rigid, thick, lignified, nonstretchable secondary cell walls. there are 2 types of sclerenchyma cells:

Sclereids or stone cells we saw in lab- short; variable shape; form hard layers such as the shells of nuts and seed coats; produce the gritty texture of pears. Fibres- long, slender; occur in strands or bundles; tiny cavity or lumen; the different hardnesses of fibres are used to make coarse rope, linen cloth, etc.

DIRECT SHOOT ORGANOGENESIS FROM LEAF EXPLANTS OF CHRYSANTHEMUM

Shoot regeneration from leaf and stem explants of 19 Chrysanthemum cultivars has been studied. Adventitious shoots could be regenerated from leaf and stem explants of 16 cultivars. Most cultivars showed the best shoot regeneration on Murashige and Skoog medium supplemented with 8.9 μM BAP and 2.7 μM NAA (0.5 mg/l). The number of regenerated shoots per explant was more than 10 in some of the cultivars studied. It has not been posssible to regenerate shoots from leaf explants of cultivars 170 and 260. The cultivars 140 and 8010 produced the highest number of shoots. Regenerated shoots can easily be rooted on half strength Murashige and Skoog medium supplemented with 0.98 μM IBA (0.2mg/l). After potting in soil and transfer to the glasshouse normal plants develop. Apparently no somaclonal variation has been observed among the regenerants.

Biolistic particle bombardment is used to transfer desirable genes into difficult to transform plants, but research is lacking on factors that can affect transformation success. This study was conducted to determine if manipulating culture conditions forRhododendron 'PJM Hybrid' and 'Iridon' chrysanthemum leaf explants could alter transient gene expression of β-glucuronidase (GUS) after bombardment.

Both rhododendron and chrysanthemum leaf explants were bombarded with 1.0 μm gold particles coated with 2 μg of linear plasmid DNA, containing the GUS gene. Different explant treatments before bombardment were tested, including time in culture on regeneration medium, growing in the presence or absence of 0, 0.09, 0.2, 0.4, or 0.6 M sucrose, and growing in the dark for 0, 3, or 6 days.

Transient GUS expression was histochemically quantified 2 days after bombardment as the percentages of GUS positive explants and the number of blue spots per GUS positive explant. Transient GUS expression was highest when 'PJM Hybrid' explants were incubated on regeneration medium 9 to 12 days before bombardment, whereas transient GUS expression by 'Iridon' explants was relatively high (>42%) for leaves incubated 3 to 18 days before bombardment.

Exposure of 'PJM Hybrid' explants to media containing different concentrations of sucrose decreased transient GUS expression by 71% compared to those grown on sucrose-free medium. Alternatively, 27% more 'Iridon' explants transiently expressed GUS compared to those on sucrose-free medium, and the mean number of spots per GUS positive explant increased from 5 to 13.

Transient GUS expression increased almost three fold when 'PJM Hybrid' explants were placed in the dark for six days compared those in the light. In contrast, the dark treatment had inconsistent effects on transient GUS expression by 'Iridon' explants. Manipulating explant culture conditions before biolistic bombardment influenced transient gene expression and illustrated that genotype-specific differences affect transformation potential.

DIRECT AND INDIRECT SHOOT ORGANOGENESIS

Orenia fournieri is also called the wishbone flower, bluewings, or torenia. It is easy to obtain and to grow.

GENERAL CONSIDERATIONS

Growth of Plants

Torenia can be grown from shot cuttings or seed. They should grow for a minimum of 8 weeks after germination. They should start flowering 12 weeks after germination. They are very vulnerable to white fly infestation.

SHOOT ORGANOGENESIS FROM PETUNIA LEAVES

The garden petunia is a popular plant is called a "cultigen" because this hybrid started in domestication. It is a hybrid of 3-4 petunia species. Garden petunias are easy to tissue culture.

GENERAL CONSIDERATIONS

Growth of Plants

Different petunia cultivars and genotypes will have different needs for PGRs, but almost all will produce shoots in vitro. Choose cultivars of the Grandiflora types. Avoid red flowering cultivars and mixed cultivars.

Eight weeks before starting the experiment, plants seeds in a soilless medium. Do not use overhead watering, as this can cause fungal and bacterial contamination. White flies can be a problem as well. It's interesting that white flies. They were a problem as well. Seedlings should be growing rosettes when explanting. Flowering plants can be used, but may be contaminated. In each petri dish, place two leaf explants adaxial side up (upper surface up).

NON-ZYGOTIC EMBRYOGENESIS

EMBRYOGENESIS

Somatic embryogenesis is the process of a single cell or group of cells initiating the developmental pathway that leads to reproducible regeneration of nonzygotic embryos capable of germinating to form complete plants. Under natural conditions this pathway is not normally followed but from tissue cultures somatic embryogenesis occurs more often and as an alternative to organogenesis for regeneration of whole plants. According to somatic embryogenesis is, initiated either by preembryogenic determined cells or by induced embryogenic determined cells. In preembryogenic determined cells, the embryogenic pathway is pre determined and the cells appear to only wait for the synthesis of an inducer (or removal of an inhibitor) to resume independent mitotic divisions in order to express their potential. Such cells are found in embryonic tissues (including scutellum of cereals), certain tissues of young in vitro grown plantlets, the nucellus, and the embryo sac (within ovules of mature plants).

Induced embryogenic determined cells, on the other hand, require redetermination to the embryogenic state by exposure to specific growth regulators such as 2, 4-D. These cells are dedifferentiated generally in microspore (anther) cultures and callus cultures. Once the embryogenic state has been reached both cell types proliferate in a similar manner as embryogenic determined cells.

Plant lets are then directly produced by following the full embryogenic pathway as a coordinated group of embryogenic determined cells. Some times individual cell or cells from the group may escape and give rise to either embryoids or nodular embryogenic callus (*e.g.* scutellar callus) consisting of proembryoids. These are embryolike structures which are bipolar units and germinate into full plantlets under suitable culture conditions. In dicotyledonous plants totipotent embryogenic cells have been most commonly obtained from explants of embryonic or young seedling tissues. Excised small tissues from young inflorescences (before maturation of floral primordia) are equally effective for the induction of somatic embryogenesis in cultures. Other explants used are the scutellum, young roots, petioles, immature leaf, and immature hypocotyl.

A high yield of embryogenic calli can also be obtained from isolated fully differentiated mesophyll cells or protoplasts in a defined culture medium. Somatic embryos germinate in situ or when they are excised and cultured individually on a fresh semisolid medium. The essential requirements for the induction and promotion of somatic embryos are primarily established by the use of suspension cultures. The in vitro development of somatic embryos was first observed in carrot suspension cultures was also able to induce somatic embryogenesis in a callus cultured on a semisolid medium. The presence of auxin in the medium is generally essential for embryo initiation. Tissues or calli maintained continuously in an auxin free medium generally do not form embryos; therefore, somatic embryogenesis is achieved in two steps.

First, the callus is initiated and multiplied on a medium rich in auxin (2, 4-D, 0.5 mg 1-1) which induces differentiation of localized groups of meristematic cells called embryogenic clumps. Second, the embryogenic clumps develop into mature embryos when transferred to a medium with a very low level of auxin (0.01-0.1 mg,-1) or no auxin at all. Hence a medium with auxin is called a proliferation medium and one without auxin an embryo development medium. Mature embryos develop in embryo development medium. Mostly 2, 4- D has been used to induce somatic embryogenesis. A substantial amount of nitrogen, usually in reduced form such as ammonium salts, is required for both embryo initiation and maturation. The source of nitrogen can be in the form of complex addenda such as coconut milk, casein hydrolysate, a mixture of amino acids or single amino acid, and ammonium ions.

MS medium contains high levels of nitrogen in the form of ammonium nitrate. Other items, such as potassium, dissolved oxygen, activated charcoal are also found to playa role in somatic embryogenesis. In monocotyledonous plants, most of which are agriculturally and medicinally important, the vegetative parts do not readily proliferate in cultures. Hence explants are best taken from embryogenic or meristematic tissues (young inflorescences and leaves).

Germination of the somatic embryo can occur only when it is mature enough to have functional shoot and root apices capable of meristematic growth. High auxin levels can inhibit development and growth of the shoot meristem and often embryos mature when the embryogenic cell suspension is transferred to a medium lacking auxin. The addition of low level of cytokinin (zeatin 0.1 μm) in combination with ABA may prove beneficial for embryogenesis of low density cell cultures. Various physical factors such as temperature also affect embryo maturation, depending on the requirement of the species. Maturation of somatic embryos proceeds more normally in complete darkness although light seems essential for somatic embryogenesis in some cultures. Somatic embryos germinate on agar medium without growth regulators. Single embryos may profit by the inclusion of low levels of zeatin (0.1 μm) in the medium.

Sometimes loss of morphogenic potential in embryogenic cultures is seen. This may be due to many reasons. Nuclear changes such as polyploidy, aneuploidy and chromosomal mutations in cultured cells may be responsible for the loss of organogenic or embryogenic potential in prolonged cultures. Altered hormonal balance within the cells or tissues may also be associated with decline in the embryogenic potential. Sometimes the non embryogenic cells of the explant increase under conditions favourable to their growth, resulting in a gradual loss of the embryogenic component during repeated subcultures.

SOMATIC EMBRYOGENESIS FROM SEEDS OF MELON

Scoop out the seeds from a ripe melon and put them into a wire mesh sieve, then with running water over the seeds rub them gently against the mesh, using it to loosen and remove the stringy fibres. Next place the cleaned seeds in a bowl of water, stir it a few times. Some seeds will float to the top....these are immature or sterile melon seeds, they are hollow and/or light-weight and will float to the top of the water. Skim away these bad seeds and discard them. Stir a few more times and repeat the process until no more sterile seeds float to the the top. Drain the water from the remaining seeds. Afterwards, line a heavy plate or baking pan with waxed paper, spread the seeds out in a single layer onto the waxed paper and place it in sunny spot to air-dry.

Stir the seeds occasionally during the next few hours to make sure all sides are exposed to fresh air, this facilitates even drying. After a day in the sun bring the seeds into the house where they continue to dry for another week or two, stir them daily so they dry evenly. If you've got rainy weather the increased humidity can prolong the drying process another week or so. Melons have thick seeds so be sure they are thoroughly dry before packing them for storage.

IMPROVEMENT OF NUTRIENT QUALITY

Transgenic crops with improved nutritional quality have already been produced by introducing genes involved in the metabolism of vitamins, minerals and amino acids. A transgenic Arabidopsis thaliana that can produce ten-fold higher vitamin E (alpha-tocopherol) than the native plant has been developed. The biochemical machinery to produce a compound close in structure to alpha-tocopherol is present in A. thaliana. A gene that can finally produce alpha-tocopherol is also present, but is not expressed.

This dormant gene was activated by inserting a regulatory gene from a bacterium which resulted in an efficient production of vitamin E. Glycinin is a lysine-rich protein of soybean and the gene encoding glycinin has been introduced into rice and successfully expressed. The transgenic rice plants produced glycinin with high contents of lysine. Using genetic engineering Prof. Potrykus and Dr. Peter Beyer have developed rice which is enriched in pro-vitamin A by introducing three genes involved in the biosynthetic pathway for carotenoid, the precursor for vitamin A. The aim was to help millions of people who suffer from night blindness due to Vitamin A deficiency, especially whose staple diet is rice. The presence of beta-carotene in the rice gives a characteristic yellow/orange colour, hence this pro-vitamin A enriched rice is named as Golden Rice. The genetic engineering is also being used to improve the taste of food, *e.g.*, a protein 'monellin' isolated from an African plant (Dioscorephyllum cumminsii) is about 100,000 sweeter than sucrose on molar basis. Monellin gene has been introduced into tomato and lettuce plants to improve their taste.

Improvement of Seed Protein Quality

The nutritional quality of cereals and legumes has been improved by using biotechnological methods. Two genetic engineering approaches have been used to improve the seed protein quality. In the first case, a transgene (*e.g.*, gene for protein containing sulphur rich amino acids) was introduced into pea plant (which is deficient in methionine and cysteine, but rich in lysine) under the control of seed-specific promoter. In the second approach, the endogenous genes are modified so as to increase the essential amino acids like lysine in the seed proteins of cereals.

These transgenic routes have helped to improve the essential amino acids contents in the seed storage proteins of a number of crop plants. *E.g.*, overproduction of lysine by de-regulation. The four essential amino acids namely lysine, methionine, threonine, and isoleucine are produced from a non-essential amino acid aspartic acid. The formation of lysine is regulated by feed back inhibition of the enzymes aspartokinase (AK) and dihydrodipicolinate synthase (DHDPS). The lysine feedback- insensitive genes encoding the enzymes AK and DHDPS have been respectively isolated from E. Coli and Cornynebacterium. After doing appropriate genetic manipulations, these genes were introduced into soybean and canola plants. The transgenic plants so produced had high quantities of lysine.

Diagnostic and Therapeutic Proteins

Experiments are going on to use transgenic plants in diagnostics for detecting human diseases and therapeutics for curing human and animal diseases. Several metabolites and compounds are already being produced in transgenic plants, *e.g.*, the monoclonal antibodies, blood plasma proteins, peptide hormones, cytokinins, etc. The use of plants for commercial production of antibodies, referred to as plantbodies, is a novel approach in biotechnology. The first successful production of a functional antibody, namely a mouse immunoglobulin IgGI in plants, was reported in 1989.

This was achieved by developing two transgenic tobacco plants-one synthesising heavy chain gamma- chain and other light kappa- chain, and crossing them to generate progeny that can produce an assembled functional antibody. In 1992, C.J. Amtzen and co-workers expressed hepatitis B surface antigen in tobacco to produce immunologically active ingredients via genetic engineering of plants. Several other therapeutic proteins have also been produced like haemoglobin and erythropoietin in tobacco plants, lactoferrin in potato, trypsin inhivitor in maize, etc. The first proteins/enzymes that were produced in transgenic plants (maize) are avidin and beta-glucuronidase and are used in diagnostic kits.

Edible Vaccines

Crop plants offer cost-effective bioreactors to express antigens which can be used as edible vaccines. The approach is to isolate genes encoding antigenic proteins from the pathogens and then expressing them in plants. Such transgenic plants or their tissues producing antigens can be eaten for vaccination/immunisation (edible vaccines). The expression of such antigenic proteins in crops like banana and tomato are useful for immunisation of humans since banana and tomato fruits can be eaten raw. Transgenic plants (tomato, potato) have been developed for expressing antigens derived from animal viruses *e.g.*, rabies virus, herpes virus. In 1990, the first report of the production of edible vaccine (a, surface protein from Streptococcus) in tobacco at 0.02 per cent of total leaf protein level was published in the form of a patent application under the International Patent Cooperation Treaty. The first clinical trials in humans, using a plant derived vaccine were conducted in 1997 and were met with limited success. This involved the ingestion of transgenic potatoes with a toxin of E. coli causing diarrhea.

The process of making of edible vaccines involves the incorporation of a plasmid carrying the antigen gene and an antibiotic resistance gene, into the bacterial cells, *e.g.* Agrobacterium tumefaciens. The small pieces of potato leaves are exposed to an antibiotic which can kill the cells that lack the new genes. The surviving cells with altered genes multiply and form a callus. This callus is allowed to grow and subsequently transferred to soil to form a complete plant. In about a few weeks, the plants bear potatoes with antigen vaccines. The bacteria E.coli, V. cholerae cause acute watery diarrhea by colonising the small intestine and by producing toxins. Chloera toxin (CT) is very similar to E.Coli toxin. The CT has two subunits, A and B. Attempt was made to produce edible vaccine by expressing heat labile enterotoxin (CT-B) in tobacco and potato.

Another strategy adopted to produce a plant-based vaccine, is to infect the plants with recombinant virus carrying the desired antigen that is fused to viral coat protein. The infected plants are reported to produce the desired fusion protein in large amounts in a short duration. The technique involves either placing the gene downstream a subgenomic promoter, or fusing the gene with capsid protein that coats the virus.

Advantages of Edible Vaccines

The edible vaccines produced in transgenic plants will sole the storage problems, will ensure easy delivery system by feeding and will have low cost as compared to the recombinant vaccines produced by bacterial fermentation. Vaccinating people against dreadful diseases like cholera and hepatitis B, by feeding them banana, tomato, and vaccinating animals against important diseases will be an interesting development.

Biodegradable Plastics

Polythenes and plastics are one of the major environmental hazards. Efforts are on to explore the possibility of using transgenic plants for biodegradable plastics. Transgenic plants can be used as factories to produce biodegradable plastics like polyhydroxy butyrate or PHB. Genetically engineered *Arabidopisis* plants can produce PHB globules exclusively in their chloroplasts without effecting plant growth and development. The large-scale production of PHB can easily be achieved in plants like *Populus*, where PHB can be extracted from leaves.

Molecular Breeding

The term molecular breeding is frequently used to represent the breeding methods that are coupled with genetic engineering techniques. Up till now, conventional breeding methods have been used to meet the food demands of the growing world population and the challenges of poverty and improved crop production and yields. However in the years to come, the development in the agriculture yields and techniques is going to be due to the use of molecular breeding programme. Linkage analysis which deals with the studies to correlate the link between the molecular marker and a desired trait is an important aspect of molecular breeding programme. In the past, linkage analysis was carried out by use of isoenzymes and the associated polymorphisms. Now a days, molecular markers are being used.

Molecular breeding involves breeding using molecular (nucleic acid) markers. A molecular marker is a DNA sequence in the genome which can be located and identified therefore molecular markers can be used to identify particular locations in the genome. Due to mutations, insertions, deletions, etc., the base composition at a particular location may be different in different plants. These differences, termed *polymorphisms*, allow DNA markers to be mapped in a genetic linkage group.

Generally, there are three types of markers used in screening/selection:

1. Morphological marker based on visible character (phenotypic expression) *e.g.*, flower colour, seed colour, height, leaf shapes, etc. Morphological markers could be dominant or recessive. There are certain constraints in using these markers as the morphological markers are easily influenced by environmental factors and thus may not represent the desired genetic variation. Some of the visible markers have not much role to play in the plant breeding programme.
2. *Biochemical marker:* The proteins produced by gene expression are also used as markers in plant breeding programmes. The most commonly used are isozymes, the different molecular forms of the same enzyme. Each individual variety has its own isozyme variability (profiles) which can be detected by electrophoresis on starch gel.
3. Molecular marker based on DNA polymorphism detected by DNA probes or amplified products of PCR, *e.g.* Restriction fragment

length polymorphism (RFLP), Randomly Amplified polymorphic DNA (RAPD), variable Number Tandom Repeats (VNTR), Microsatellites, etc. Plant breeders always prefer to detect the gene as molecular marker, although it is not always possible. Molecular markers provide a true representation of the genetic make up at the DNA level. They are consistent and free from environmental factors, and can be detected much before the development of plants occur. The advantage with a molecular marker is that a plant breeder can select a suitable marker for the desired trait which can be detected well in advance. A large number of markers can be generated as per the needs.

The molecular markers to be used in plant breeding programme should have the following characteristics:

- The marker should be closely linked with the desired trait,
- The marker screening methods should be effective, efficient, reproducible and easy to carry out,
- The entire analysis should be cost effective.

Molecular makers are of two types:

1. Based on nucleic acid (DNA) hybridisation- This involves the cloning of the DNA piece followed by the hybridisation with the genomic DNA, which is later detected. The Restriction fragment length polymorphism (RFLP) was the very first technology employed for the detection of polymorphism, based on the DNA sequence differences. RFLP is mainly based on the altered restriction enzyme sites, as a result of mutations and recombinations of genomic DNA. The procedure involves the isolation of genomic DNA and it's digestion by restriction enzymes. The fragments are separated by electrophoresis and finally hybridised by incubating with cloned and labelled probes.
2. Molecular markers based on PCR amplification. Polymerase chain reaction (PCR) is a novel technique for the amplification of selected regions of DNA. The most important advantage is that even a minute quantity of DNA can be amplified and the PCR- based molecular markers require only a small quantity of DNA to start with. Random amplified polymorphic DNA (RAPD) markers use PCR amplification where the DNA is isolated from the genome and is denatured. The template molecules are annealed with primers and amplified by PCR. The amplified products are separated on electrophoresis and identified. Based on the nucleotide alterations in the genome, the polymorphisms of amplified DNA sequences differ which can be identified as bends on gel electrophoresis. Amplified fragment length polymorphism (AFLP) is a novel technique involving a combination of RFLP and RAPD. AFLP is

based on the principle of generation of DNA fragments using restriction enzymes and oligonucleotide adaptors (or linkers), and their amplification by PCR.

Microsatellites

Microsatellites are the tandemly repeated multiple copies of mono-, di-, tri-, and tetra nucleotide motifs. In some instances, there are unique flanking sequences present in the repeat sequences. Primers are designed for such flanking sequences to detect the sequence tagged microsatellites (STMS) which is done by PCR.

Commercial Use of Transgenic Plants

The main goal of producing transgenic plants is to increase the productivity. In 1995-96, transgenic potato and cotton plants were used commercially for the first time in USA. By the year 1998-99, five other major transgenic crops cotton, maize, canola, soybean, and potato were introduced to the farmers. These accounted for about 75 per cent of the total area planted by crops in USA. There are still a lot of concerns regarding the harmful environmental and hazardous health effects of transgenic plants. The major areas of public concern are- the development of resistance genes in insects, generation of a super weeds by mutation, etc. Certain other legal and regulatory hurdles pertaining to commercial use of transgenic plants, needs to be addressed.

Bioethics in Plant genetic Engineering There are issues and concerns regarding the use of transgenic crops and their effects on the health and the environment in general.

The major concerns about GM crops and GM foods are:

- Effect of GM crops on biodiversity and environment- As the GM crops are created artificially, there is no natural process of evolution in their development. Hence, there is a question of this affecting the biodiversity and overall effect on the environment.
- The risk of transfer of transgene from GM crops to pathogenic microbes- Antibiotic marker genes are used to identify and select the modified cells. If GM food containing antibiotic resistance marker gene is consumed by animals and humans, there is a risk that the transgene will transfer from GM food to microflora of human and animals. This may lead to the gut microbes to become resistant to antibiotics.
- The transfer of genes from animals into Gm crops for molecular farming may change the fundamental vegetable nature of plants.
- The GM crops may bring about changes in evolutionary patterns. The plants adapt to the changing environment in the natural way

by changing their genes and developing better races with superior traits which ultimately leads to the development of evolved races and varieties. What will be the evolutionary pattern of the GM crops? There are concerns about the effect of transgene flow from GM crops to other non-GM plants and the alteration of these non-GM crops.

- There is a risk of transferring allergens (usually glycoproteins) from GM food to human and animals.
- There is a risk of "gene pollution" *i.e.,* transfer of transgene of GM crop through pollen grains to related plant species and development of super weeds.
- There are also some religious issues related to the consumption of transgenic plants with animal genes introduced into them, especially, for some strict vegetarian people and some ethnic groups with certain food preferences and restrictions.
- There is a need to study thoroughly as to how the genetically engineered plants will affect the ecological balance, once they are released in the environment.

3

Cell Culture of Plants: Methods and Techniques

Plant tissue culture encompasses culturing of plant parts on an artificial medium. The plant parts can be a single cell, tissue or an organ. It is also referred to as micropropagation. Plant tissue culture was practically implemented for the first time by Haberlandt, a German scientist, in 1902. Later in 1934, Gautheret found successful results on in-vitro culture of plants. The basic key used in plant tissue culture is the totipotency of plant cells, meaning that each plant cell has the potential to regenerate into a complete plant. With this characteristic, plant tissue culture is used to produce genetically identical plants (clones) in the absence of fertilization, pollination or seeds.

Plant tissue culture encompasses culturing of plant parts on an artificial medium. The plant parts can be a single cell, tissue or an organ. It is also referred to as micropropagation. Plant tissue culture was practically implemented for the first time by Haberlandt, a German scientist, in 1902. Later in 1934, Gautheret found successful results on in-vitro culture of plants. The basic key used in plant tissue culture is the totipotency of plant cells, meaning that each plant cell has the potential to regenerate into a complete plant. With this characteristic, plant tissue culture is used to produce genetically identical plants (clones) in the absence of fertilization, pollination or seeds.

METHODS AND TECHNIQUES

Raupp *et al.* evaluated the relative yield, product quality and soil life after long-term (17 years) organic or mineral fertilization. They found that production yields varied: yields of potatoes and rye were lower and yields of spring wheat were similar in the organic system. However, their evaluation also showed that humus content and biological activity in the soil were greater, that products from the organic system had a better storage quality and that vegetables had lower nitrate contents. Apart from the quality of the product from different production systems, the production potential of organic in comparison with conventional production systems is important. Research findings seem contradictory as production yields are found to be lower, equivalent, or higher.

Simulating the production potential after long term differing management systems showed the following: Droogers *et al.* compared the production potential of two farming systems (biodynamic and conventional) by converting 'static' basic soil properties into a 'dynamic' assessment using simulation modelling. Soil conditions on two farms - one having been managed biodynamically for 70 years - were investigated with morphological and physical methods. A simulation model including 30 years climatic data was used to predict water-limited potato yields, showing that the simulated yields were significantly higher on the biodynamic fields.

RHIZOSPHERE - INTERACTION WITH PLANT GROWTH

No published research was identified in New Zealand on this topic. The international work published emphasised the influence that soil flora and fauna have on plant growth in an organic system. Scullion *et al.*, for example, described how management before conversion can have an impact on soil fungi, that in turn will influence plant growth under the organic system. Yeates *et al.*,described soil microbe and faunal diversity in Welsh soils, and while the soil types may differ, the message to enhance diversity is the same. This paper also described indicators that could be used to define soil "health" in an organic system. Eason et al.and Cook *et al.*,both described how management can impact on soil microbe and fungi populations and how this in turn can influence plant growth.

TRACE ELEMENTS

Condron et al presented a comparison of soil quality under conventional and organic management in New Zealand. One area of focus in the paper was on trace elements in soils under organic and conventional pasture and their effects on animal health. Trace element impacts on animal health were also mentioned in some of the more generic animal health papers described earlier. Other than this work, there was very little information on trace elements in soils under organic pastures.

MICROBIAL ACTIVITIES AND BIOMASS

Synthetic water soluble fertilizers are known to be detrimental to soil microorganisms and therefore to influence biological and physical soil characteristics negatively in the long term. Mader et al.studied the effect different fertilization intensities on different crops have on soil microorganisms at different soil depths. They compared organic and conventional farming systems in Switzerland and reported that soil microbial biomass was significantly higher under the biodynamic system. Wood reported on studies in the UK, that showed an increase in earthworm populations with additions of manure; and comparisons of conventional and biodynamic farms in New Zealand showed that the biodynamically farmed soils had better structure, lower bulk density, higher organic matter content and respiration rates, and higher earthworm populations. Pfiffner et al described a long-term trial studying the effect of different farming systems (biodynamic, organic and conventional) on ground beetles.

The number of species was consistently higher (193 per cent) in the biodynamic than in the conventional (100 per cent) and organic (188 per cent) plots. The number of species differed from 18-24 in biodynamic plots, compared to 19-22 in organic and 13-16 in conventional plots. In 1998 Pfiffner *et al.*, reported on a long-term trial into the effect of different farming systems (biodynamic, organic and conventional) on earthworm populations. The earthworm biomass and density, and the number of juveniles and earthworm species were significantly higher in the biological than the conventional or organic plots.

Ryan reviewed studies with regard to the question 'Is an enhanced soil biological community a consistent feature of alternative agricultural systems?' Studies that examined four groups of soil organisms, comparing the soil biological community in conventional, organic and biodynamic management system were reviewed, and a case study of biodynamic and conventional dairy farms in Australia included. Ryan's conclusion was that the enhanced soil community found on the biodynamic farms relative to conventional neighbours should not be considered as a definitive feature of alternative agricultural systems, but rather as an effect of the higher input of organic matter.

Lytton-Hitchins et al.came to the same conclusion after comparing the physical and chemical properties of biodynamically and conventionally managed pastures in NE Victoria, Australia. The more favourable properties found on the biodynamically managed soils are attributed to decreased grasing pressure, longer intervals between irrigations, reduced tractor traffic and intermittent applications of compost and hornmanure preparations.

Ryan *et al.*, undertook a glasshouse experiment with soil samples from 3 biodynamic (without conventional fertilizers for 17 years) and conventional (regular inputs of soluble P and N fertilizers) dairy pastures. Plant nutrient uptake was examined by assessing the response of white clover, perennial rye grass and indigenous VAM fungi to the addition of 4 levels of soluble P and N. Their findings indicated that the soils had not developed substantially different processes to enhance plant nutrient uptake. The experiments did not examine the response of plant nutrient uptake to non-water-soluble fertilizer, which might have produced different results.

DEVELOPMENT OF PLANT CELL

The plant cell after formation normally increases very considerably in size. This process may be studied in two situations:

- When the cells are in intact organs,
- When they are in isolated systems.

The first indicates the scope and characteristics of the process, the second the factors that control the several aspects of it. The two groups of studies based on the different situations complement each other, and it is therefore proposed to consider below, first, observations made with intact organs, and secondly, observations made with isolated fragments. This treatment of plant cell growth is not intended to be exhaustive and one topic in particular is omitted.

For two reasons no discussion is attempted of the nature of the auxin effect in cell growth. The subject is omitted, since it has recently been discussed in some detail by Audus and, secondly, since it is possible that an elucidation of this problem can only be developed after the nature of cell growth has been more fully explored. Hitherto plant cell growth has frequently been described as cell extension. In our view this term is an unfortunate one. It is probably a legacy of the period when cell growth was considered a simple matter of an inflation due to the absorption of water. Recent work, however, has shown that the increase in size of the cell involves increases in many components of the system, and we therefore prefer the term cell growth to the older one of cell extension.

PROCESS OF CELL GROWTH

The process of cell growth involves the changes that occur between the meristematic and the mature fully vacuolated state. The meristematic cell is small isodiametric and without a prominent central vacuole. As the cell enlarges a central vacuole develops and the shape changes.

Eventually a large parenchymatous cell is established with a thick wall and a thin peripheral layer of cytoplasm. This set of changes is readily studied on the root. It is probable that this is a more satisfactory experimental object than the coleoptile, since with this organ the earlier stages of growth are not readily accessible to investigation.

The root, being an organ of unlimited growth, has an apical meristem from which cells are being continuously generated. These cells immediately they are formed commence enlarging, and thus at increasing distances from the apex the tissue is composed of cells in progressively advanced stages of development. A similar situation is involved in the apex of the stem. Here, however, the position is complicated by the development of lateral members from the surface of the meristematic dome. In the root the surfaces of the growing regions are not covered with developing primordia, and this circumstance has been exploited in the development of two sets of techniques for the study of cell growth. Since the surface of the growing region is naked in the root direct observations may be made microscopically on epidermal cells.

The dimensions of epidermal cells may be measured at increasing distances from the apex and the growth of these cells determined over the whole range of development. This technique has been used in principle by Burstrom, Brumfield and by Goodwin. It has been shown that sliding growth is not involved, and changes in surface cells may therefore be taken as representative of corresponding changes occurring in deeper tissues. This general technique has yielded important results, and it has the merit that it involves observations on a more or less uniform tissue. It carries the limitation, however, that only a limited variety of observations can be made with it. It cannot, for instance, yield data regarding metabolic conditions in developing cells. In order to avoid this difficulty another experimental approach has been proposed by Brown and Broadbent.

These workers cut successive sections from the apex backwards, made observations on each section, and related the values obtained to the total number of cells in each section. They used a technique for determining the number of cells in a tissue originally developed by Brown and Rickless. The range of observations that can be made with this technique is clearly greater than that possible with the earlier procedure. It involves, however, two possible sources of error. Cell division does not cease at the same distance from the apex in all regions of the root, and all cells in the section are therefore not exactly at the same stage of development.

Secondly, sectioning necessarily involves a destruction of cells the extent of which cannot easily be estimated. The first source of error is not decisive, since the distance over which cell division ceases is small relative to the total distance over which growth occurs, and the average given by the section therefore represents approximately the position for all the cells in the section. The error from the destruction of cells can be reduced by increasing the length of the section and thus decreasing the proportion of damaged cells. These data are from the observations of Brown and Broadbent. The full development occurs over a distance of about 5.0 mm. from the apex and involves about a twenty- to thirtyfold increase in volume.

The enlargement is accompanied by important metabolic changes. In the present connexion it may be noted that the enlargement in the early stages is accompanied by the gradual suppression of the capacity to divide. The zone that stretches from the tip to about 1.5 mm. behind it is that over which mitotic figures are normally distributed. At the same time the observations of Wagner and of Gray and Scholes have shown that the mitotic frequency is greatest at the tip and decreases progressively with increasing distance from it. Brown has emphasized that the frequency of mitotic figures may not be an index of the relative rate of division. In this case, however, since the frequency decreases to zero it probably is, and the data therefore suggest that after formation at the apex the increase in the average cell size is accompanied by a corresponding increase in the average length of the interphase. As growth proceeds the probability of division decreases, and at a certain point the probability becomes so small as to constitute virtual cessation of the process. The conditions that determine this transition have not been explored, but it is clear that the fact that it occurs indicates that growth in the early stages is accompanied by metabolic changes that tend to suppress division.

At about the point at which division ceases a prominent central vacuole appears in the cell. This vacuole, which is probably a development from a dispersed vacuolar system in the meristematic cell, increases in size with the enlargement of the cell. During growth, however, the percentage water content is increasing, indicating that the volume of the vacuole is increasing more rapidly than the other components, and that the enlargement of the cell is being determined primarily by the absorption of water. At the same time it is evident that the increase in water content is being sustained by relatively smaller increases in other phases of the system.

These data indicate that the dry weight of the cell increases about tenfold during the course of growth. Preston and Clark and Wirth have shown that the cellulose content of cells in the coleoptile increases during growth and Wirth has demonstrated corresponding changes with soluble carbohydrates. The increase in dry weight of the cell in the root is therefore probably due partly to similar changes.

It is evident, however, that it is not due only to changes in these constituents. In 1941 Blank and Frey-Wyssling described a series of important observations on cell growth in the coleoptile of maize. They reported an increase in protein content during the enlargement of the cell. Blank and Frey-Wyssling found similar changes occurred during cell enlargement in the hypanthium of *Oenothera*. Kopp calculated that in the root the protein content increased, and Brown and Broadbent demonstrated this condition experimentally in pea roots.

The appropriate curve is reproduced from the data of Brown and Broadbent, and it is evident from this that protein increases progressively during growth of the cell, resulting in at least a fivefold increase. At the same time it may be noted that when growth ceases there is a slight but significant decrease in protein, the final level being lower than it is when growth stops, although greater than it is in the meristematic region. The increase in protein necessarily implies an increase in the general metabolic activity of the cell during growth.

Kopp calculated and Brown and Broadbent later demonstrated an increase in respiration rate per cell during growth. The data of Brown and Broadbent suggest, however, that when growth ceases there is a slight fall in respiration which is coincident with the decrease in protein. The increase in protein content might also be expected to increase the quantities of particular enzymes. We have recently been examining the changes in the activities of certain enzymes during the growth of the cell. These data have been obtained by taking successive sections along bean roots, killing the sections by freezing at - 10°C., determining on the killed tissue the activities of certain enzymes, and finally reducing the values for each section to a unit cell basis. The activities of three enzyme systems have been investigated: a dipeptidase which hydrolyses alanylglycine, an invertase, and an acid phosphatase.

The activity of the dipeptidase at increasing distances from the apex of the root of barley has already been examined by Bottelier, Holter and Linderstrom-Lang, and Wanner and Leupold have determined the relative sucrose-splitting capacity in different regions of the root of maize. In the bean roots growing in the conditions of our experiments growth continues over about the first 8 mm. The data show that over this range there is about a fivefold increase in dipeptidase, about a twofold increase in phosphatase, and about a twentyfold increase in invertase activity.

Further, it is evident that in about the region where growth in length is ceasing the activities of the respective enzyme systems decrease. Determinations of protein in cells of bean roots are not available, but in view of the earlier observations with peas it is probable that the increase in enzyme activity is due to the differentiation of part of the protein increment into enzyme bodies, and that the decrease in enzyme is similarly partly due to a decrease in protein. It is an important aspect of the situation that the activities of the three enzymes do not change relatively to the same extent. The relative increases are different and the ratios of the activities of any two of the enzymes therefore change during the course of growth.

A similar position is indicated by some observations of Linderstrom-Lang and Holter on the activities of what may be two distinct enzymes cleaving leucyl-glycine and alanylglycine respectively at increasing distances from the tip of the root of barley. The data, although they are not expressed in terms of unit cell number, nevertheless show that the ratio between the two enzymes changes as the distance increases. Clearly since the ratios between the enzyme activities change it is evident that the metabolic pattern must change during the course of growth.

During the growth of the cell a differentiation probably occurs in the metabolic state such that the relative intensities of different reactions change in the course of the process. In particular, these observations on the enzyme complement suggest that the metabolic activities of the young cell are different from those of the mature. This conclusion is consistent with the results of an investigation by Morgan of the amino-acid composition of the proteins in different parts of the growing zone of the bean root. The proteins from sections taken at increasing distances from the apex of the root have been hydrolysed and the hydrolysates examined chromatographically. It is evident that the composition of the proteins varies considerably. Such differences in the composition of the proteins suggest of course corresponding differences in the generalized metabolic pattern.

This general conclusion is also consistent with a variety of observations by other workers. Berry and Brock have reported that the respiration of the tip of the onion root is more sensitive to cyanide than is that of the mature tissue, and Kopp has found that per unit protein it is lower at the tip than it is in the mature regions.

The data of Brown and Broadbent suggest that the increase in protein per unit increase in volume is greater in the earlier than it is in the later stages of growth. Finally, Burstrom has shown that the reactions of root cells to sugar, temperature and heteroauxin are different in the early and late stages. The growth of plant cells is interesting on account of the fact that the protoplasm is covered with a solid wall. Any irreversible increase of the cell size involves the growth in area of this wall, which consists partly of crystallized substances. The development of cell walls, especially in the cells of higher plants, passes through two different stages.

As long as there is growth in area, they remain thin and flexible. This fine membrane is called the *primary cell wall*, and it is said to grow in area by *intussusception*. As soon as the final shape and size of the cell are reached, the membrane is strengthened by adding new layers to the primary wall. During this second stage the cell wall grows in thickness by *apposition*. The combined layers are called the *secondary cell wall*.

The thickness of this is often so large that the slender primary wall is overlooked or neglected, especially in materials of technical importance such as wood pulp, textile fibres, etc. Indirect methods of investigation, such as polarized light and X-rays, indicate that the primary wall consists of a loose submicroscopic network of crystallized strands of cellulose or chitin, arranged in a dispersed texture.

In the secondary wall the framework is much more compact. This compactness is obtained by a parallel arrangement of the submicroscopic strands forming a highly anisotropic *parallel texture*. Two different types of growth in the area of the primary wall can be distinguished:

TIP GROWTH

Tip growth of filamentous cells such as fungal hyphae, hairs (*e.g.* root hairs), latex tubes, pollen tubes, etc. In this case, only the tip of the cell is involved in growth, and no intussusception occurs away from the end. Sometimes the tip growth of very long cells (e.g bast fibres of ramie, which reach a length of over 20 cm.) continue after the formation of the secondary wall has already started in the middle body of the cell.

EXTENSION GROWTH OF CELLS IN ELONGATING TISSUES

Extension growth of the cells in elongating tissues. In this case the whole length of the primary wall is said to increase in area by intussusception. There is an enormous bibliography on this subject, because the extension growth is stimulated by auxin, which was considered at the time as the hormone of the extension growth. However, the question how auxins intervene in this process cannot be solved until the mechanism of the growth in area has been cleared up. For a long time it was thought that the extension growth consisted of simple water intake, and the turgor stretch of the cell wall caused thereby.

But it has been shown that this type of cell wall growth is much more complicated, since it involves respiration, biosynthesis, and morphogenesis. Moreover the work needed for the elastic deformation of the growing cell wall is negligible as compared with the total energy available for growth. As extension growth concerns tissues with numerous cells which expand simultaneously at different rates, the geometrical problem how contiguous polyhedra can change their shape without being individualized arises. Krabbe thought that a disintegration of the tissue was inevitable and suggested the hypothesis of *sliding growth.*

This implies that the cells of a tissue assume the individuality of independent unities growing along each other like unicellular protophytes. This is an extreme interpretation of sliding growth it is true, but it demonstrates clearly that such an idea is theoretically unsound. Priestley published other arguments against the possibility of this type of growth; sliding of the two adjacent cell walls would interrupt the channels of pits, so that their orifices would no longer correspond.

In consequence one might expect to find a large number of unilateral pits in tissues in which sliding growth had occurred. As this is not the case, Priestley postulated that the adjacent cell faces must grow together at the same rate, and he termed this *symplastic growth.* His theory cannot, however, explain how latex tubes and other specialized cells grow across tissues and thereby come in touch with cells which previously were not their neighbours. This frequently occurs when derivatives of the cambial cells, such as tracheids and fibres, elongate in the tissue.

Two adjacent cells are pushed apart by the tip of a cell which intrudes between them. This has been termed intrusive growth by Sinnot and Bloch or interposition growth by Schoch-Bodmer. The last author emphasizes that only the tips of elongating fibres behave in this way. According to her, the walls of the central body of the cell do not grow any longer, so that the entity of the tissue is secured. Pits are only found in that part of the cell wall, and the fibre tips are free from them. From this description it follows that different parts of the cell wall must behave differently. A plant cell does not grow as a whole, but in a differentiating cell there are growing and full-grown parts at the same time.

CELL CULTURE

Plant tissue culture encompasses culturing of plant parts on an artificial medium. The plant parts can be a single cell, tissue or an organ. It is also referred to as micropropagation. Plant tissue culture was practically implemented for the first time by Haberlandt, a German scientist, in 1902. Later in 1934, Gautheret found successful results on in-vitro culture of plants. The basic key used in plant tissue culture is the totipotency of plant cells, meaning that each plant cell has the potential to regenerate into a complete plant. With this characteristic, plant tissue culture is used to produce genetically identical plants (clones) in the absence of fertilization, pollination or seeds.

METHODS AND TECHNIQUES OF PLANT TISSUE CULTURE

In plant tissue culture, plants or explants such as pieces of leave, stem or root is cultured in a specific plant medium, which contains essential plant nutrients and hormones. Other plant growth factors like light and temperature are maintained and regulated by using artificial conditions. All the procedures of plant tissue culture are conducted under sterile (aseptic) conditions. The explants then develop stem, roots and leaves. The generated plantlets are hardened before planting in outdoor conditions. To start with plant tissue culture, one should wipe his/her hands, forceps and other equipments with alcohol or other sterilizer to prevent microbial contamination. Following this, the explants are surface sterilized by using chemical solutions such as bleach or alcohol. Though mercury chloride is an effective sterilizer, it is rarely used due to its potential toxicity. After sterilization, the explants are introduced into a plant medium, which can be either solid or liquid.

The cells divide and differentiate into plant parts, thus giving rise to a complete plant. Speaking about the plant medium, liquid type is prepared by mixing inorganic salts, organic nutrients (sucrose), vitamins, minerals and plant growth regulators (auxin and cytokinin). In addition to these ingredients, solid medium contains gelling agent (agar). Nutrient and plant hormones amount vary, depending on the objective of plant tissue culture. For example, in order to induce more roots, auxin amount should be high. Plant tissue culture techniques include the culture of protoplast (a cell without cell wall), meristem, node, anther, ovule, embryo and seed.

APPLICATIONS OF PLANT TISSUE CULTURE

Plant tissue culture has wide applications in agriculture. One major advantage of plant tissue culture is the production of disease and/or pest resistant varieties, thus indirectly increasing the crop yield. Commercially, it is used directly for the propagation of plants that are hard to propagate in natural conditions. Horticultural plants such as orchids, roses, banana, strawberries, potatoes, apples, etc. are successfully cultured in in-vitro conditions. Plants with valuable secondary products are grown in the controlled conditions by using plant tissue culture technique. It is also used for the propagation of medicinal herbs on a large scale. With plant tissue culture, it is possible to generate virus-free plantlet of vegetatively propagated plants. In experimental biology such as plant breeding, cell biology, biotechnology and genetics, plant tissue culture is applied in order to solve plant related problems.

It allows screening of cells for the desirable characters such as early fruit bearing, disease resistance and drought resistance. Another important application is generation of a novel hybrid by crossing two distantly related species having advantageous traits. In-vitro fertilization and/or pollination of plants is possible, irrespective of the hindrances in natural conditions. Overall, plant tissue culture is used for the conservation of germplasm.

THE PRODUCTION OF CELLS

Moving fluids such as waterfalls, fountains, and the wakes of boats can generate specific forms. The study of the motion of fluid particles and the shape changes that the fluids undergo is called kinematics. The ideas and numerical methods used to study these fluid forms are useful for characterizing meristematic growth. In both cases, an unchanging form is produced even though it is composed of moving and changing elements. An example of an unchanging form composed of changing and displaced elements in plants is the hypocotyl hook of a dicot, such as the common bean. As the bean seedling emerges from the seed coat, the apical end of the hypocotyl bends back on itself to form a hook.

The hook is thought to protect the seedling apex from damage during growth through the soil. During seedling growth (in soil or dim light) the hook migrates up the stem, from the hypocotyl into the epicotyl and then to the first and second internodes, but the form of the hook remains constant. The hooked form is maintained over time, while different tissues first curve and then straighten as they are displaced from the seedling apex during growth. If a mark is placed at a fixed point on the surface, it will be displaced (indicated by the arrow), appearing to flow through the hook over time. If we mark a specific epidermal cell on the seedling stem located close to the seedling apex, we can watch it as it flows into the hook summit, then down into the straight region below the hook. The mark is not crawling over the plant surface, of course; plant cells are cemented together and do not experience much relative motion during development. The change in position of the mark relative to the hook implies that the hook is composed of a procession of tissue elements, each of which first curves and then straightens as it is displaced from the plant apex during growth. The steady form is produced by a parade of changing cells. A root tip is another example of a steady form composed of changing tissue elements. Here, too, the form is observed to be steady only when distance is measured from the root tip. A region of cell division occupies perhaps 2 mm of the root tip. The elongation zone extends for about 10 mm behind the root tip. Phloem differentiation is first observed beginning at 3 mm from the tip, and functional xylem elements may be seen at about 12 mm from the tip.

A marked cell near the tip will seem to flow first through the region of cell division, then through the elongation zone and into the region of xylem differentiation, and so on. This shifting implies that developing tissue elements first divide and elongate, and then differentiate.

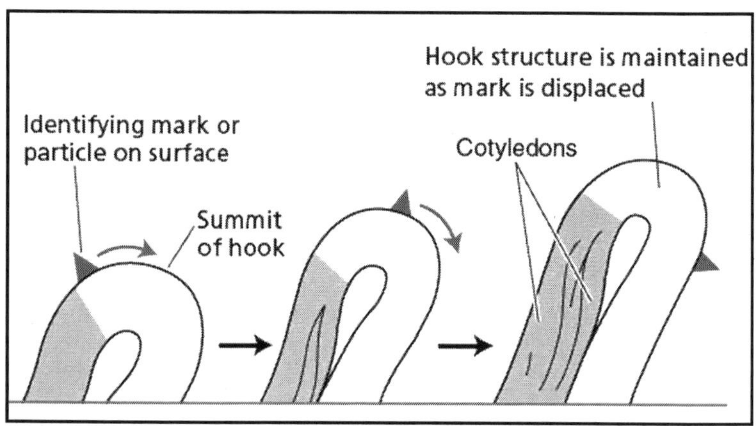

Fig. The dicot Hypocotyl hook is an example of a Constant form Composed of changing Elements.

In an analogous fashion, the shoot bears a succession of leaves of different developmental stages. During a period of 24 hours, a leaf may grow to the same size, shape, and biochemical composition that its neighbour had a day earlier. Thus, shoot form is also produced by a parade of changing elements that can be analyzed with kinematics. Such an analysis is not merely descriptive; it permits calculations of the growth and biosynthetic rates of individual tissue elements (cells) within a dynamic structure.

TISSUE ELEMENTS ARE DISPLACED DURING EXPANSION

As we have seen, growth in shoots and roots is localized in regions at the tips of these organs. Regions with expanding tissue are called growth zones.

With time, meristems move away from the plant base by the growth of the cells in the growth zone. If successive marks are placed on the stem or root, the distance between the marks will change, depending on where they are within the growth zone. In addition, all of these marks will move away from the tip of the root or shoot, but their rate of movement will differ depending on their distance from the tip.

From another perspective, if you were to stand at the tip of a root that had marks placed at intervals along the axis, you would see that all marks would move farther away from you with time. The reason is that discrete regions on the plant axis experience displacement as well as expansion during growth and development. As a given region of the plant axis moves away from the apex, its growth velocity increases (the rate of elongation accelerates) until a constant limiting velocity is reached equal to the overall organ extension rate. The reason for this increase in growth velocity is that with time, progressively more tissue is located between the moving particle and the apex, and progressively more cells are expanding, so the particle is displaced more and more rapidly. In a rapidly growing maize root, a tissue element takes about 8 hours to move from 2 mm (the end of the meristematic zone) to 12 mm (the end of the elongation zone).

Beyond the growth zone, elements do not separate; neighbouring elements have the same velocity (expressed as the change in distance from the tip per unit of time), and the rate at which particles are displaced from the tip is the same as the rate at which the tip moves through the soil. The root tip of maize is pushed through the soil at 3 mm h^{-1}. This is also the rate at which the nongrowing region recedes from the apex, and it is equal to the final slope of the growth trajectory.

THE GROWTH OF VELOCITY

The velocities of different tissue elements are plotted against their distance from the apex to give the spatial pattern of growth velocity, or growth velocity profile. Growth velocity increases with position in the growth zone. A constant value is obtained at the base of the growth zone. The final growth velocity is the final, constant slope of the growth trajectory equal to the elongation rate of the organ. In the rapidly growing maize root, the growth velocity is 1 mm h^{-1} at 4 mm, and it reaches its final value of nearly 3 mm h^{-1} at 12 mm.

(A) The growth velocity profile plots the velocity of movement away from the tip of points at different distances from the tip. This tells us that growth velocity increases with distance from the tip until it reaches a uniform velocity equal to the rate of elongation of the root. (B) The relative elemental growth rate tells us the rate of expansion of any particular point on the root. It is the most useful measure for the physiologist because it tells us where the most rapidly expanding regions are located.

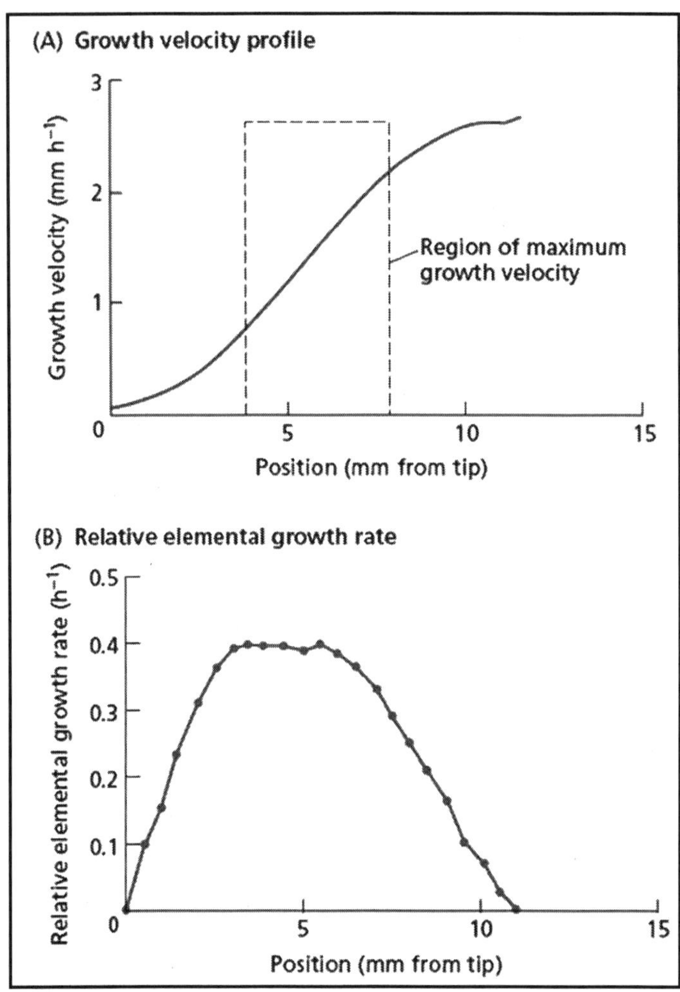

Fig. The Growth of the Primary root of Zea mays (maize) can be represented Kinematically by two related growth curves.

If the growth velocity is known, the relative elemental growth rate, which represents the fractional change in length per unit time, can be calculated. The relative elemental growth rate shows the location and magnitude of the extension rate and can be used to quantify the effects of environmental variation on the growth pattern.

The Relative Elemental Growth Rate

One way to characterize the growth pattern along an axis mathematically is to plot the distance of a tissue element from the apex versus time. The resulting curve is called the growth trajectory. The slope of the curve at any given point is equivalent to the growth velocity at that region of the axis. With the apex as the reference point, the growth trajectory is concave. Marked tissue elements at different locations may be followed simultaneously to generate a family of growth trajectories.

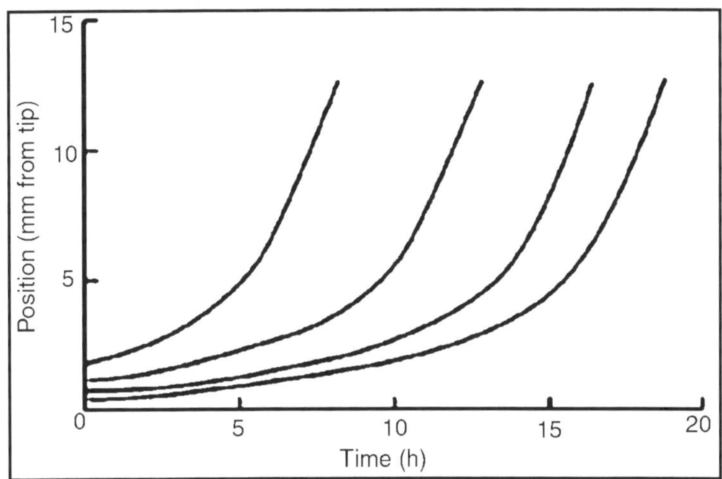

Fig. Growth Trajectories of *Zea mays* (maize): Growth Trajectories plot the positions of points spaced along the root axis as a Function of time.

Note that distance is measured from the root tip. This is termed a material specification of growth because real or material particles are followed.

As the regions of the plant axis move away from the apex, their growth velocity increases (*i.e.*, the cells accelerate) until a constant limiting velocity is reached equal to the overall organ extension rate. The reason for this increase in growth velocity is that with time, progressively more tissue is located between the moving particle and the apex, and progressively more cells are expanding, so the particle is displaced more and more rapidly.

In a rapidly growing maize root, a tissue element takes about 8 hours to move from 2 mm (the end of the meristem) to 12 mm (the end of the elongation zone). The root tissue element accelerates through this region. Beyond the growth zone, elements do not separate; neighbouring elements have the same velocity, and the rate at which particles are displaced from the tip is the same as the rate at which the tip is pushed away from the soil surface.

The root tip of maize is pushed through the soil at 3 mm h^{-1}. This is also the rate at which the nongrowing region recedes from the apex, and it is equal to the final slope of the growth trajectory. The growth trajectory is a material description of growth because an individual element is followed through time. The growth trajectory provides a way to convert spatial information into a developmental time course.

For example, if the amount of suberin in the root increases with position, the growth trajectory will give information on the time required for the observed accumulation of suberin in cells moving between any two spatial points. To characterize the local expansion rate, we must consider the growth velocities at the apical and basal ends of a small segment of tissue. If the apical end of the segment is moving faster than the basal end, the segment is elongating.

The velocity difference, divided by the segment length, gives the relative growth rate of the segment. Now, imagine that the segment shrinks to a point x. To find the relative growth rate, r, at point x, we can use calculus: The value of r is given by the velocity gradient, dv/dx, where v is the growth velocity. This measure of the local growth rate is called the relative elemental growth rate. It represents the fractional change in length per unit time and has units of h^{-1}.

If the growth velocity is known, the relative elemental growth rate can be evaluated by differentiation of the growth velocity with respect to position. Or we can approximate r by measuring the relative growth rates of closely spaced segments of initial length L. For each segment, $r = (1/L)(dL/dt)$.

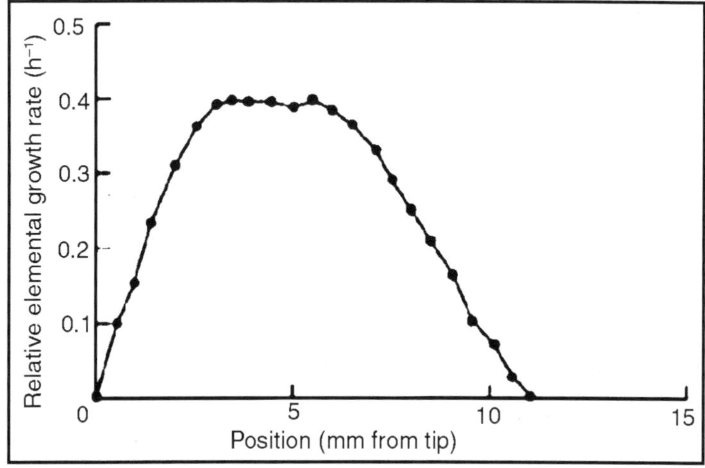

Fig. Relative Elemental Growth rate of *Zea mays* (maize).

The relative elemental growth rate tells us the rate of expansion of any particular point on the root. It is the most useful measure of root growth rate for the physiologist because it tells where the most rapidly expanding regions are located. The relative elemental growth rate profile, $r(x)$, is a spatial description of the expansion pattern in the growth zone. In the maize root, the relative elemental growth rate has a maximum of 0.40 h^{-1} at 4 mm and drops to zero by 12 mm.

That is, the tissue element at 4 mm is expanding 40% h^{-1}, while tissue beyond 12 mm has ceased expansion. Again, to understand the profile in physiological terms, we must consider that individual cells expand more and more rapidly as they are displaced to 4 mm and then less and less rapidly as they are displaced through the basal part of the growth zone. The relative elemental growth rate profile shows the location and magnitude of the extension rate and can be used to quantify the effects of environmental variation on the growth pattern. It is also the basis for studying the physiology of growth. The graph of $r(x)$ can be compared to the distribution pattern of any substance thought to be regulating the growth rate. A close correlation between the two curves would provide circumstantial evidence for the role of the substance as a limiting factor for growth.

Embryo Maturation Requires Specific Gene Expression

The Arabidopsis embryo enters dormancy after it has generated about 20,000 cells. Dormancy is brought about by the loss of water and a general shutting down of gene transcription and protein synthesis—not only in the embryo, but also throughout the seed. To adapt the cell to the special conditions of dormancy, specific gene expression is required. For example, the *A*bscisic *A*cid *I*nsensitive3 (*ABI3*) and *FUSCA3* genes are necessary for the initiation of dormancy and are sensitive to the hormone abscisic acid, which is the signalling molecule that initiates seed and embryo dormancy.

ABI3 also controls the expression of genes encoding the storage proteins that are deposited in the cotyledons during the maturation phase of embryogenesis.The *LE*AFY *C*OTYLEDON*1* (*LEC1*) gene also is active in late embryogenesis. Because *lec1* mutants cannot survive desiccation and do not enter dormancy, the embryos die unless they are rescued through isolation before desiccation occurs. The rescued embryos will germinate in culture and produce fertile plants, which are like wild-type plants except that they lack the 7S storage protein and they have leaflike cotyledons with trichomes on their upper surface.

The normal appearance and development of the mature *lec1* mutants indicates that the *LEC1* gene is required only during embryogenesis. Although the most obvious defects of the *lec1* mutants are seen only in the maturation phase embryo, mRNA from *LEC1* gene expression can be detected throughout embryogenesis. It has been proposed that *LEC1* is a general repressor of vegetative development and its expression is necessary throughout embryogenesis.

STRUCTURE OF CELL WALL

The basic unit in the structural organization of the cellulosic framework of the plant cell wall is the cellulose molecule. Cellulose molecules are not units of definite molecular weight but long chains of varying length formed by the condensation of at least a 100-120 and possibly many times this number of /3 d-glucose molecules1. Studies of cellulose walls with the X-ray and polarized light have furnished evidence that the molecules of cellulose are aggregated into bundles known as micelles.

The micelles have been estimated upon the basis of X-ray photographs to have a diameter of approximately 5 or 6 mp and a length of about 60 m. A unit of this size would consist of about sixty parallel cellulose chains each being made up of about 120 glucose units. It is probable that in many cell walls long chain-like molecules formed by the condensation of other sugars are associated with the cellulose chains in the micelles. Originally the micelles were believed to be well defined units which were cemented together by some non-crystalline material.

Recent work indicates however, that many of the cellulose chains are much longer than the micellar aggregates and extend from one micelle to another, thus welding the micelles into a coherent an astomosing system. It is also probable that the micellar aggregates vary in size. Cellulose walls are no longer regarded, therefore as a system of discrete crystalline units bound together by some cementing substance but rather as a structure composed of molecular aggregates are welded together by interlocking cellulose molecules. There seems to be little doubt regarding the presence of intermicellar spaces. The spaces form an interconnecting system between the anastomosing micelles.

Black areas represent intermicellar spaces; white areas represent the cellulose micelles:

(A) Cross section,

(B) Longitudinal section.

In walls that are almost pure cellulose, such as the secondary wall of cotton fibres, it is possible that the intermic-ellar spaces are filled chiefly with water.The smallest visible units of cellulose walls are delicate thread like strands or fibrils. In primary walls these fibrils form a loose anastomosing network the meshes of which are usually filled with colloidal pectic compounds.

In secondary walls the fibrils are often grouped into coarser strands which wind around the cell in a steep spiral the angle of which may vary indifferent layers and even in different parts of the same layer.

PHYSICAL PROPERTIES OF CELL WALLS

The physical properties of primary cell walls differ from those of secondary walls. These differences may be ascribed to the greater abundance of cellulose in secondary walls and to differences in the structural organization of the cellulose in the two kinds of walls. Both primary and secondary walls are transparent to wavelengths of the visible spectrum and both are usually quite pcrmeable to most substances dissolved in water that compare favourably with those of steel.

The breaking strength of a flax fibre, for example, may be as great as 110 kg. per mm.2 of wall area while the tensile strength of a hardened spring steel ranges between 150 and I 70 kg. per mm2. Although the tensile strength of primary walls is considerably less than that of secondary walls it is only rarely that they are subjected to strains in excess of their breaking strength. Primary cell walls are usually quite elastic.

The mesophyll cells of the leaves of some species, for example, are known to undergo reversible changes in volume of 30 per cent or more in response A to changes in turgor pressure. Secondary walls, on the other hand, are elastic only within very narrow limits and break suddenly when their tensile strength is exceeded. The elasticity of primary walls is undoubtedly related to the loose open mesh work of cellulose strands that compose it, while the close parallel orientation of the cellulose micelles in secondary walls prevents any appreciable elongation without rupture.

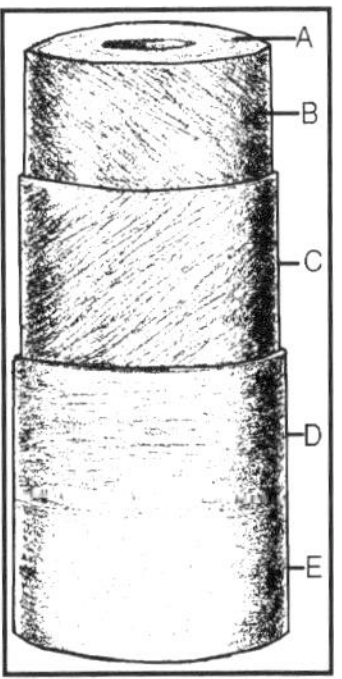

Fig. Diagram Illustrating the Structure of a Thickened Plant Cell Wall.

CHEMICAL CONSTITUENTS OF CELL WALLS

Cellulose is by far the most abundant compound found in the cell walls of the higher plants. Associated with cellulose in all the primary cell walls of vascular plants are greater or lesser amounts of pectic compounds. Their presence in the middle lamella and in the intermicellar spaces of the primary walls has already been noted. Lignin is an important constituent of the walls of most of the cells that make up woody tissues and it also occurs commonly in other thickened walls.

Although lignin can be isolated from cell walls by suitable treatments as a brownish amorphous substance it is probably altered considerably in the process. Its structural formula and molecular weight are unknown. Since the lignins produced in different species appear to differ in their chemical properties it seems very probable that lignin is not a compound of definite molecular weight but a complex mixture of a number of chemically similar substances. The lignin of spruce wood has been extensively studied and several workers have suggested that its molecular weight lies between 800 and 900.

Lignin first appears in the middle lamella and primary wall and later canbe detected in the secondary wall. When associated with cellulose, lignin is present in the spaces between the micelles and not within the micelles. Lignifled walls are usually freely permeable to water and solutes. The tensile strength of lignified cell walls is the same as that of cellulose walls but lignified walls resist compression better than cellulose walls. The increased resistance of lignified walls to compression is explained by the assumption that the presence of lignin in the intermicellar spaces welds the cellulose micelles into a single coherent mass and thus prevents the bending and buckling of the cellulose strands when they are subjected to compression strains.

Cutin is the name applied to the mixture of wax-like materials found on the outer surface of the epidermal cell walls of leaves, stems, fruits, and other organs. These wax-like substances are intimately associated with cellulose and often with pectic substances producing a wall of great structural complexity. Cutinized cell walls are relatively impermeable to water. The presence of cutin in the outer walls of epidermal cells greatly reduces the evapouration of water from the surfaces of plant tissues. Subcrin is similar in many of its properties to cutin. It constitutes an important part of cork cell walls and it is also found in the walls of a few other specialized types of cells. Most of the surface of perennial plants, aside from the leaves and very young stems is covered with suberized cell walls.

Such walls are relatively impermeable to water. Hemicell-uloses are a poorly defined group of poly saccharides associated with cellulose in plant cell walls. They are not chemically related to cellulose, as the name implies, but possess very different chemical and physical properties. Callose is the name given to a carbohydrate membrane substance found in the perforated septa (sieve plates) of the sieve tubes. Similar material presumably of the same chemical composition, has been found in pollen grains and constitutes the inner layer of pollen tubes. It has also been reported as occurring in the fungi. The exact chemical composition of callose is unknown since it has never been obtained in sufficient amounts to permit quantitative determinations. Chitin, a nitrogenous substance common in the exoskeleton of insects, is a constituent of the walls of many fungi and bacteria. It has been reported as being present in the walls of certain algae but it is unknown in any of the higher plants. Tannins are commonly found in the cell sap but they also occur in the walls of certain tissues, especially cork and wood cells.

Mucilages are common constituents of the outer walls of many water plants and occur also in the outer walls of some seed coats, in glandular hairs and in other specialized tissues. Inorganic compounds such as silica and salts of calcium, iron, and other metals are also present in some plant cell walls. None of these inorganic compounds, however, are regarded as essential constituents of the cell wall.

PROTOPLASM

In most mature cells of the vascular plants protoplasmis present only as a thin layer covering the inner surface of the cell walls but in some specialized cells branching strands of protoplasm also extend across the vacuole. Under high magnification the protoplasm of active cells appears as a colourless fluid, in which are suspended numerous tiny granules and droplets of insoluble materials. These granules frequently exhibit active Brownian movement. The fluid component of protoplasm also is frequently in motion, streaming around the inner surfaces of the cell walls. Embedded in this fluid component, and carried passively by it, are the specialized bodies known as plastids.

Although protoplasm appears to be a simple liquid, no simple liquid could possibly possess the remarkable powers of synthesis, assimilation, reproduction growth and sensitivity that characterize the protoplasm of living plant cells.The properties and behaviour of protoplasm clearly show that it is not a substance but that it must be regarded as a complex system of substances. This system is dynamic; it is constantly undergoing changes yet at the same time the changes are so regulated and controlled that the system is not disrupted. A cell is alive only so long as the organization of this dynamic protoplasmic system is maintained.

The protoplasmic system of living plant cells almost invariably contains a well differentiated globular body known as the nucleus. All of the protoplasm outside of the nucleus is designated as cytoplasm. Although the nucleus is separated structurally from the cytoplasm by a delicate membrane the two components are not physiologically isolated. The nucleus and the cytoplasm appear to be mutually dependent and there is good reason to believe that the nucleus controls and regulates the physiological processes that occur in the cytoplasm. Probably because of the close relationship between the nucleus and the cytoplasm many biologists use the terms protoplasm and cytoplasm synonymously.

THE CHEMICAL COMPOSITION OF PROTOPLASM

Since protoplasm is a dynamic system of substances it is not possible to subject it to chemical analysis without destroying it. In the strict sense, therefore, it is impossible to discover the chemical composition of protoplasm. It is possible, however, to examine the substances present after the protoplasmic system has been destroyed and to determine their chemical composition and relative abundance. A number of such studies have been made.

Water is the chief component of all physiologically active plant protoplasm usually making up more than 90 per cent of the system. The water content of the protoplasm of dry seeds, on the other hand, may be less than 10 per cent. Most attempts to determine the chemical composition of plant protoplasm have been made upon species of the myxomycetes. At certain stages in their life history these organisms consist of naked masses of labile protoplasm. They are often found flowing over rotten logs in damp woods. The fact that the myxomycetes provide relatively large quantities of protoplasm entirely free from cell wall material has made them a favourite object for chemical analysis. Even in such organisms, however, not all of the constituents of the plant body can be regarded as integral parts of the protoplasm. Distributed throughout the protopl-asmic mass are particles of foods and other inert materials which cannot be separated from the protoplasm. As shown in this analysis proteins and other nitrogen-containing compounds constitute the bulk of the organic matter in the plasmodium of this species. Many different varieties of proteins are known to occur in the protoplasm of plant cells. They are compounds of enormous molecular weight and undoubtedly make up a large proportion of the labile structural framework of the protoplasm.

Lipids constitute a smaller fraction of the protoplasm than the proteins. Three types of lipids occur in the protoplasm; the true fats (oils), the phosphatides (phospholipids) and the sterols, of which phytosterol is an example. The oils are generally suspended in the protoplasm in the form of minute globules. They are probably more important as food reserves than as actual constituents of the protoplasm. The phospho-lipids and sterols, on the other hand, are believed to be essential constituents of the protoplasmic system. The water-soluble carbohydrates, amino acids, etc., present in the plasmodium of this species are probably almost entirely foods.

The inorganic compounds (mineral matter) in plant cells are chiefly the phosphates, chlorides sulfates, and carbonates of magnesium, potassium, sodium, and calcium. It supplies no more information regarding the organiz-ation of protoplasm, however, than a chemical analysis of the ground up debris of a wrecked house would furnish regarding the structure of that house. This point is worthy of emphasis, since with the progress of biology it becomes clearer and clearer that the properties of protoplasm are as much a function of its physiochemical organization as of the specific kinds of compounds present.

Although we commonly refer to protoplasm as the essential constituent of all living cells, it is evident that there are at least as many different varieties of protoplasm as there are species of plants and animals. All of them, however, are dynamic systems composed of the same types of compounds, and all of them possess a colloidal organization of a complex type.

METHODS OF SINGLE CELL ISOLATION

FROM PLANT ORGAN

The most suitable material for the isolation of single cells is the leaf tissue since a more or less homogeneous population of cells in the leaves offer good candidates for raising defined and controlled large scale cell culture. From such intact plant organs single cells can be isolated using mechanical or enzymatic methods.

Mechanical Method

Mechanical isolation involves tearing or chopping surface sterilised explant to expose the cells followed by scrapping of the cells with a fine scalpel to liberate the single cells hoping that it remained undamaged. Gnanam and Kulandaivelu (1969) developed a procedure to isolate mesophyll cells, active in photosynthesis and respiration, from mature leaves of several species of dicot and monocot. The procedure involves mild maceration of 10g leaves in 40 ml of grinding medium (20 µ mol sucrose, 10µ mol Mgcl2, 20 µ mol tris – HCL buffer, Ph7.8) with a mortar and pestle. The homogenate is passed through two layer of muslin cloth and cell thus released are washed by centrifugation at low speed using the same medium. The mechanical isolation of free parenchymatous cells can also be achieved on a large scale.

Enzyme Method

Takebe et al. (1968) treated tobacco leaf tissue with enzyme pectinase and obtained a large number of metabolically active cells. The potassium dextran sulphate in the enzyme mixture improved the yield of free cells. Isolation of single cells by the enzymatic method has been found convenient, as it is possible to obtain high yields from preparation of spongy parenchyma with minimum damage or injury to the cells.

This can be accomplished by providing osmotic protection to the cells while the enzyme macerozyme degrades the middle lamella and cell wall of the pranchymatous tissue. Applying enzymatic method to cereals (Hordeum vulgare, Zea mays) has proven rather difficult since the mesophyll cells of these plants are apparently elongated with number of interlocking constrictions, thereby preventing their isolation.

From Cultured Tissue

The most widely applied approach is to obtain a single cell system from cultured tissue. Freshly cut piece from surface sterilised plant organs are simply placed on a nutrient medium consisting of a suitable proportion of auxins and cytokinins to initiate cultures. Explant on such a medium exhibit callusing at the cut ends, which gradually extends to the entire surface of the tissue.

The callus is separated from an explant and transferred to a fresh medium of the same composition to enable it to built up a mass tissue. Repeated sub-culture on an agar medium improves the friability of the callus, a pre-requisite for raising a fine cell suspension in a liquid medium. The piece of undifferentiated and friable callus are transferred in a continuously agitated liquid medium consisting of flask or suitable vials. Agitation is done by placing the culture medium exerts mild pressure on small pieces of tissue, breaking them into free cells and small cell aggregates. Further, it augments the gases exchange between the culture medium and the culture air and also ensures uniform distribution of cells as wells as clumps in the medium.

METHODS OF TISSUE CULTURE STUDY

Process of biotechnology in tissue culture may be broadly placed in four groups.

Somatic Cell Fusion

This is achieved by inducing two or more pro top lasts to fuse and the fusion product is nurtured to produce a hybrid plant. Though, this phenomenon is of common occurrence, protoplasts of some plant species fail to fuse. Some scientists have shown the production of hybrids that cannot be produced by the conventional hybridisation methods because of-sexual or physical incompatibility among certain plants.

Genetic Engineering

Experiments are being conducted by some scientists to introduce nuclei, chloroplasts, viruses, DNA, mitochondria, plastids, etc., keeping in view the capability of isolated protoplast to ingest "foreign" bodies by a process similar to the process of endocytosis often seen in cells of protozoan and other animals.

Wall Biosynthesis

Scientist have made use of isolated protoplast for the study of wall biosynthesis and deposition because of the ability of cultured protoplasts to regenerate a cell wall rapidly.

Study of Protoplast Population

As the protoplasts can be manipulated in a manner similar to that of the micro-organism, selection of mutant cell lines and the cloning of cell population is possible by using microbiological methods.

FUNCTION OF TISSUE CULTURE

In the recent years plant protoplast, cell and tissue cultures have become an important tool for crop improvement, commercial production of natural compounds and in the development of forestry.

Some of the areas of biotechnological application of cultured plant protoplasts/cells/tissues are:

- Tissue culture applications in order to capitalise on the totipotency of cells,
- Cells and protoplast cultures coupled 'with DNA vectors to overcome problems caused by barriers to gene transfer through sexual means,
- Culture of plant cells for the production of useful compounds,

- Extension and increase of deficiency of biological nitrogen fixation, and
- Transfer of genes for nitrogen fixation ability to non-fixing species.

Due to the manipulation of plant tissues in the laboratory this technique has been referred to by some researchers, as a 'botanical laser' whose numerous uses are yet to be fully understood.

APPLICATION IN AGRICULTURE

The major areas of application of tissue culture are briefly highlighted here.

Improvement of Hybrids

Development of cell fusion and hybridisation techniques have solved the problem of incompatibility of plants and widened the scope of production of new varieties with in a short time. For example breeders obtained somatic hybrid of wild and cultivated potatoes (*S. tuberosum, S. chacoense*) and succeeded in the induction or organogenesis.

The somatic hybrid plant inherited many characters, *viz.*, intermediate leaf morphology, stomata, forms and colour of tubers prolonged flowering, large and fertile pollen grains, high yield, resistance against Y-virus. However, haploid plant materials available as protoplast, cell and tissue culture systems are currently being evaluated for the use in transfer of foreign genetic material to selected plant species by protoplast fusion, transformation, transduction and organelle transfer.

Production of Encapsulated Seeds

The two 'techniques, somatic embryogenesis and organogenesis are the alternative methods for regeneration of the whole plant from cultured tissue *in vitro*. Moreover, the recent works on the wrapping an *in vitro* derived embryos in a seed coat, *i.e.*, production of synthetic seeds are important because it may improve agriculture due to its low cost.

Production of Disease Resistant Plants

Many plant species, which propagate vegetatively are systematically infected by virus, bacteria, fungi and nematodes. Their inoculum is carried over several generations resulting in continued adverse effect of productivity and quality of crops. In order to ensure highest possible yield and quality, it is necessary to provide disease free stock plants to growers. Tissue culture techniques have solved this problem and minimized the time of biological testing.

Unless large scale population of pure inoculum of test pathogens are available, it is difficult to pursue the establishment of pathogenicity and crop loss-assessment as it is done in field conditions. Now it has become possible to carry out such experiments in laboratory within short span of time by using tissue culture technologies. Scientists have discussed the following advantages for the study of several aspects of host-pathogen interactions and responses.

They are:

- Ability to isolate host cells without wounding,
- Control of inoculum of pathogen and number of host cells,
- Ability t'J change the nature of host pathogen interaction by altering the constituent of growth medium,
- Presence of only one or a few major host cell types and
- Easy to apply and remove materials, *e.g.*, labelled precursors form cultured cells.

Production of Stress Resistance Plant

Biochemical mechanisms exist is Culture cells which determine the resistance to biocide chemicals and provide the theoretical promise for selection *in vitro*. For example: cell suspension tolerant to 2,4-D when the cells were subcultured for six months in liquid medium supplemented with increasing amount of herbicides. The cells were able to grow in one rnM (milli mol) 2,4-D, while the control suspension was completely inhibited at O.3mM 2,4-D. It is suggested that tolerance was the result of induction of enzymatic systems responsible for the degradation of 2,4-D.

Uses of protoplasts as a system to select cell lines tolerant to herbicides have not been extensively explored. Protoplast technology can be used to increase the possibility to obtain monoclonal lines and offer the opportunity for intraspecific\transfer of cytoplasmic factor of resistance to some type of herbicide. A potential application of transfer of herbicides tolerance factor is the transplant of cytoplasmic organelles, for l example chloroplasts.

Transfer of Nif Gene to Eukaryotes

Nitrogen fixing ability, a genetic character, exists in prokaryotic diazotrophs. However, one of the major tasks is the transfer of this character to eukaryotes. In recent years, researches are being done to solve this problem through tissue culture techniques coupled with the recombinants DNA technology. Historically, nitrogen fixation by rhizobia was believed to occur through symbiosis. For the first time an excitement was caused in scientific community with the discovery by Holsten *et al.* (1971). They obtained active rhizobia in the absence of nodules, leghaemoglobin and bacteroids which were apparently necessary in the intact plants. They established *Rhizobium joponicum.* In cell suspension of soybean roots. Callus induced from root explants of soybean on a specific medium was inoculated with R. joponicum. Later on it was microscopically observed that infection threads were formed by the bacteria which were present between intracellular spaces. They multiplied inside cells. Moreover, development of nitrogenase in soybean callus- Rhizobium system growing on solid was also observed and it was also found that only specific (isomorphic) cells of callus are vulnerable to infection by the bacteria. In addition to improvement the bacterial strains and increased nodulation, it is necessary to seek those genotypes with the efficient photosynthesis and improved partitioning of carbohydrates to nodules.

Future Prospects

In addition to work done successfully on nif gene transfer, there are other important genes which have been clones, for example:

- Phaseolin and leghaemoglobin genes in soyabean;
- Storage protein genes in soybean,
- Genes of ribulose bi-phosphate carboxylase/oxygenase (Ru BP case) of pea, maize, wheat, etc.

Success achieved on these aspects would certainly promote in green revolution. Moreover, improvement in primary productivity by conversion of C_3 plants into C_4 ones through genetic engineering techniques hopefully would increase the primary productivity.

APPLICATION IN HORTICULTURE AND FORESTRY

Tissue culture is applied in a number of cases in horticulture and forestry.

Micropropagation

Microprogation is a perfect alternative to asexual propagation of important plants and trees, because of following reasons:

- In this method only a small piece of tissue is needed to generate millions of clonal plants a year
- This method provides a possible alternative method for developing resistance in many species.
- It solved problem of quarantine for introduction of art disease in new areas and provide a mean for international exchange of plant material.
- By this method multiplication can be made in any season.

Regeneration of plantlets from cultured plant cells and tissues has been achieved in many trees of high economic value. Many of the studies are aimed at large scale micropropagation of important trees yielding fuel, pulp, timber, oils or fruits. Therefore, clonal forestry and horticultural are gaining an increasing recognition as an alternative for tree improvement. However, strategies for transferring cultured plants *in vitro* to field conditions are based on relatively higher priced horticultural species rather than agricultural and forestry species. At planting out stage the plantlets fail to survive because of sudden change in the environment and invasion by soil microbes. So the regenerates should be transferred first to green house and then to field. The humidity should be controlled by covering the plants with transparent polythylene sheets. Some important plants on which work is done are: *Acacia nilotica, Albizia lebback, Aprocera, Azadirachta Indica, Bauhinia purpurea, Butea monosperma, Dalbergia spp, Dendrocalmus strictus, Eucalyptus spp, Ficus religiosa, Morus spp, Populus spp, Shorea robusta, Tectona grandis, Cryptomeria japonica, Picea smithiana and Pinus spp* (all gymnosperms).

In Vitro Establishment of Mycon-biza

Mycorrhizal fungi show highest specialisation of parasitism. But major problems with them is their failure to grow on an artificial medium in laboratory. Therefore, establishment and multiplication of mycorrhizal fungi on cultured tissue of the same host plant, if successfully developed, may be a good tool for handling mycorrhizal fungi, production of high potential inoculum and their establishment in root systems of nursery plants in horticulture and forestry and the plantation of mycorrhiza infested seedlings into the field. Many attempts have been made to establish vesicul arbuscular mycorrhizal (VAM) fungi in axenic culture but unfortunately none of them got success. Moreover, micorhizal fungi have been cultured only on cortical tissues of roots which were seperated from the whole plant, as in root organ cultures where it acted as food base. VAM fungi have very high degree of specialisation for food base on root cortex.

APPLICATION IN INDUSTRIES

Production of useful compounds by cultured plant cells has become a field of special interest in various biotechnological programmes. However, much attention has been paid on the production of pharmaceutical and other secondary compounds such as essential oils, food flavourings and colourings which are used- by the fast food, ice cream and confectionary industries. This can be achieved by selection of specific cells producing high amount of desired compounds and the development of a suitable medium. In general, secondary metabolites produced by plant cell cultures are rather in small amount but strains of cells producing the same are in greater amount than those found in the intact plants have been isolated by clonal selection.

Commonly two methods are employed for the selection of specific cells: Single cell cloning and cell aggregate cloning. The difficulties associated with isolation and culture of single cells limit application of this method. The later may appear to be more time-taking but easier than the first one. Recently, *in vitro* production of high amount of useful compounds has increased with the success obtained so far in experimental studies. It is hoped that in near future, the industrial production of such compounds by using these techniques would be possible. Similarly, monogenic cultures of nematodes have been used for the study of mechanisms of action of nematicide. This technique can be used extensively in industry to supply nematodes for nematicide screaning programmes.

4

Research on Plant Movements and Development

PLANT MOVEMENTS

In response to various environmental factors, or by themselves plants exhibit various types of movements. Many of these movements are autonomous in nature and others are in response to external stimuli like light, soil, water, touch and chemicals. Any of the above mentioned agents which induce certain changes in the plants is referred to as stimulants. Reaction of the plant body to such stimulus is called reaction of the plant body to such stimulus is called response.

Neither all parts of the plants are involved in receiving the stimulus, nor do all plants respond to the given stimulus. In many cases the region that is sensitive to stimulus is called the region or perception, which is separated from the region of response; sometimes both may occur at the same site. For a proper response, the stimulus should be applied for a certain period of time below which the response will not be fully achieved. Such period is called presentation time.

The presentation time of the stimulus is provided with optimum dosage. Nonetheless, once the stimulus is provided, the plant body after absorbing the stimulus is provided, the plant body after absorbing the stimulus is provided, the plant body after absorbing the stimulus, reacts to it and responds to it. The time between the application of stimulus and response is called reaction time or latent time. If the stimulus is provided successfully without any time for relaxation, the plants start responding weakly and finally they don't show any response. This state is called Fatigue. The extreme fatigue leads to a condition called'tetanus".

Types of Movements: Movements in plants can be broadly classified into physical movements and vital movements. In the case of physical movements the physical forces act upon the non-living structures of the plant parts and as a response, these structures exhibit certain movements. On the other hand living cells of plant body are involved, in movements which are called vital movements.

Physical: Elaters of Bryophytes, dehiscence of capsules, opening and closing of peristomial teeth in Moss capsules, bursting of capsules, swelling of seeds, etc are the few examples for physical movements which respond to the changes in the moisture content. Most of the structures mentioned above are made up of lignin or such hydrophilic substrates. These substances adsorb water and swell, or lose water and shrink. This expansion and contraction brings about the movement, some times with considerable force and violence.

Vital Movements

Movements of locomotion: Autonomic type: These movements are basically the movements caused by protoplasmic activity. They are a kind of involuntary type that go on all the time. Nevertheless their movement is an active process requiring the utilisation of metabolic energy (ATP). They also respond to certain factors like temperature, respiratory inhibitors and growth hormones. Ciliary movement, amoeboid movement, gliding movement and protoplasmic streaming (cyclosis and rotation) movements are the examples of this kind.

Movements of Locomotion: Paratonic type

These are induced movements, various external factors like light, chemicals, temperature act as stimuli. Accordingly they either move towards the stimuli or move away from the stimuli. Photatactic, chemotactic and thermo tactic are the examples of this kind.

Movements of curvature: Autonomic

Autonomic: The upward and downward movement of the basal leaflets of Indian telegraph plant (Desmodium gyrense) is a non- growth movement exhibited by a rhythmic osmo-regulatory process. Nutation and circum nutation movements exhibited by stem tips and tendrils are the growth movements. Because of the alternate or unequal growth of the plant structures, they show such bending and rotatory movements.

Movements of Curvature: Paratonic type

Paratonic movements are the induced movements, which may be directional or non- directional. They may either growth movements or non-growth movements.

TROPIC MOVEMENTS

The growth movement in response to external stimuli, like light, gravitation, water and touch are called phototropism, geotropism, hydrotropism and thigmotropism respectively.

Phototropism: The bending movement of stem tips towards the light source is called phototropism. Here the light provides the stimulus. The actinic part of the white light that causes the positive curvature movement is blue light. This indicates that the plant cells which respond to this light must possess a pigment system which is capable of absorbing the blue light. This pigment has been identified as riboflavin and it is called as"cryptochrome". How this pigment brings about the differential growth, which is responsible for the growth curvature, is not very clear. Added to this the involvement of auxin in the growth/elongation of the cells is of prime importance. When the stem tips are exposed to unilateral light, it has been found that the auxin gets distributed unequally between the cells exposed to light and the cells in the unexposed area. This unequal distribution (*i.e.,* more in the unexposed area and less in the exposed cells), brings about more growth in the cells containing more auxin and relatively less growth in the cells with less auxin. As a result of this the stem bends towards light.

Whether the unequal distribution of auxin in the cells caused by sunlight is due to the lateral transportation from the illuminated area to darker area or due to inactivation of auxin found in the cells of illuminated area is a muted question. Recent experiments indicate that the active form of auxins found in the cells of the illuminated area become inactive form rather than the complete destruction. This change in active form an inactive form is very important. More over, the exposure of cells to unilateral light may also cause the synthesis of ABA locally and it may be responsible for the inhibition of growth of the cells. This may be responsible for the initiation of growth of cells. This may be achieved by extrusion of K ions from the cells of illuminated area to other side and make the cells flaccid or it may bring about a change in active form of auxin to inactive form.

A new dimension has been added to this phenomenon with the observation that the treatment of stem tips with 0.5 per cent colchicines, makes the stem tip insensitive to either light mediated or auxin mediated growth curvature. This finding again supports the idea of auxin bringing about the growth effect by activating the tubulin polymerising system. However, the exact mechanism is still not very clear though we have some facts and figures of observed growth movements.

Geotropism (Barytropism or Gravitropism): When a seeding is kept horizontally on the soil, the radicle always grows towards soil, and the plumule grows away from the soil. This type of growth movement of radicle and plumule is called positively geotropic and negatively geotropic respectively. But some runners and rhizomes grow parallel to the surface of the soil and this is referred to as Diagiotropic. When the lateral roots grow obliquely into the soil, it is referred to as plagiogeotropic phenomenon.

Sites of Geotropic Perceptions: In stem, the maximum perception is received by stem tip, but in root, it is the root cap that acts as the site of perception.

Mechanism: The analysis of auxin distribution in the plant as shown in the figure indicates that stem tip contains highest concentration and the root tip has next highest. This suggests that the optimal growth of stem tip requires higher concentration and root tip requires lower concentration for the maximal growth. On the other hand it can be stated that the concentration required for the maximum growth in stem tip is inhibitory to the growth of root tip, but the concentration that is optimal for root tip is inadequate for the maximal growth of the stem tip.

When a seedling is placed horizontally on the soil the auxin moves down by gravitational force and accumulates in cells that are found facing the soil. The stem apex and root apex respond differently to different concentration of auxins. Higher concentration in the step tip enhances the growth of the cells which have received more auxins. So the upward growth curvature occurs. On the contrary the growth in the root apex is inhibited because of higher concentration, but lower concentration in the upper region promotes the cell growth, hence the roots exhibit downward growth curvature.

This differential growth behaviour of stem apex and root apex to different concentration can be further sustained by planting the seedling in a rotating pot of the Clinostat so as to make the auxin concentration uniformly distributed. Because of this, root and stem grow straight with out any curvature. The above observations were found to be inadequate for explaining geotropic movement. Nevertheless, Berthelot and Haberlandt suggested that starch containing amyloplasts act as statoliths.

When root is placed horizontally, the statoliths because of their weight are displaced from the original position and move downwards to settle on the membranes below. This contact of statoliths with cell membrane of that side causes irritation and inhibits cell growth on that side, but other side of the cell grows normally. Unfortunately this theory failed to take into consideration of those plants which are devoid of statoliths but still showing geotropic response (Eg. Fungi etc).

Recently it has been demonstrated the root cap has a profound effect on the geotropic curvature, because the root caps synthesise IBA and release it to the lower cells. This transport of IBA causes the inhibition of cell growth in that region and other cells which are free from IBA grow normally and exhibit growth curvature. Still it is difficult to visualise how IBA brings about this effect. Nonetheless, one thing is clear that the cells in stem tip and root tips though totipotent, their potentiality to receive stimulus is different and the mechanisms they operate are also different. To achieve the growth curvature they use their own phytohormones.

Hydrotropism: Whatever may be the position or direction in which the seedling is placed in an inert vermiculite or saw dust after providing water from the base, the radicle always grows/towards water. This type of growth movement is called hydrotropism. The exact mechanism by which the root tip seeks water, as it grows towards soil, is not known except for the fact that roots have in-built potentially for absorption of water and mineral salts. It is the innate desire that drives root to grow towards water. Physiological mechanisms that generate these movements are not known.

Thigmotropism: Certain parts of the plant body like tendrils, etc., are highly sensitive to touch. These organs are found to have tactile pits for the perception sites of the stimulus. When cells come to contact with the surfaces like wood, stems, etc., get irritated and synthesise and liberate IBA. This causes the inhibition of growth of these cells at that surface while other cells which are away from the contact region continue to grow normally. Thus, they bring about the coiling of the tendrils around the supporting structure.

Paratonic Nastic Movements: The movements are a kind of induced movements, but these may be also due to growth or osmoregulation.

Photonasty: This type of movement is induced by light and darkness. Photonastic movements otherwise called Nyctinastic or sleeping movements are exhibited by the leaves of many leguminous plants where basal petioles of the leaflets have a bulbous structure called pulvinus. Here the leaflets either fold upwards or downwards. The mechanism responsible for this type of movement is believed to be due to turgour changes in response to blue part of the white light. The exact mechanism of differential turgour changes that occur in upper and lower cells of the pulvinus is not known. However, it is suspected that K ion influx and efflux is responsible for such turgour movements.

Chemonasty: Insectivorous plants like Drosera, Nepenthes, Dionea, exhibit this type of chemonastic movements. These plants grow in N2 difficient areas and require nitrogenous products for their sustenance. Hence these plants have devised various insect trapping mechanisms in which the specialised structures are well developed. When an insect, mistaking for glistening honey or attractive food comes in contact with the perception sensitive sites, the hairs which are sensitive pass on the stimulus to hinge part of the leaf, which immediately responds in closing the leaves over the prey. This excitation, conduction of excitation and folding movement is due to ions influx and efflux, which actually cause turgour movements.

Seismonasty: This is brought about by mechanical or seismonastic irritability in the force of touch. Mimosa pudica, Biophytum sensitivum and others exhibit this type of movement. The leaf pulvinae of these plants are made up a central vascular strand surrounded by a large area of cortical cells, with large intercellular spaces. The cortical cells found in the lower part of the basal pulvinus of the leaves have thinner cell walls than the upper cells. These thin walled cells are highly sensitive and collapsible.

When the terminal pinnae of the bipinnate leaves of Mimosa pudica are stimulated by touch (mechanical irritability), the stimulus is transmitted from the tip of the base at the speed of 10-20 cm/sec. This is virtually similar to that of the conduction of nerve impulse. Once the stimulus is conducted to the pulvinus, the upper thin walled cortical cells collapse by losing cell water into intercellular spaces. This results in the upward bending movement and pinnae close. When the stimulus reaches base of the leaf the cortical cells found in the lower part collapse by the same process and the entire that are released in the perception of stimulus and the chemicals that are transmitted all along the length of the xylem elements are not clearly known. However, it has been suspected that the mechanical shock or stress releases the IBA, which in turn sets on the efflux of K ions in the cortical cells, which is actually responsible for the collapsing of the cells and the movement. The leaves take 30 minutes to one hour to recover from this shock.

PLANT GROWTH IN DIFFERENT WAYS

Growth in plants is defined as an irreversible increase in volume. The largest component of plant growth is cell expansion driven by turgor pressure. During this process, cells increase in volume manyfold and become highly vacuolate. However, size is only one criterion that may be used to measure growth. Growth also can be measured in terms of change in fresh weight—that is, the weight of the living tissue—over a particular period of time. However, the fresh weight of plants growing in soil fluctuates in response to changes in water status, so this criterion may be a poor indicator of actual growth. In these situations, measurements of dry weight are often more appropriate. Cell number is a common and convenient parameter by which to measure the growth of unicellular organisms, such as the green alga *Chlamydomonas*. In multicellular plants, however, cell number can be a misleading growth measurement because cells can divide without increasing in volume.

Growth is assessed by a count of the number of cells per milliliter at increasing times after the cells are placed in fresh growth medium. Temperature, light, and nutrients provided are optimal for growth. An initial lag period during which cells may synthesize enzymes required for rapid growth is followed by a period in which cell number increases exponentially. This period of rapid growth is followed by a period of slowing growth in which the cell number increases linearly. Then comes the stationary phase, in which the cell number remains constant or even declines as nutrients are exhausted from the medium. For example, during the early stages of embryogenesis, the zygote subdivides into progressively smaller cells with no net increase in the size of the embryo.

Only after it reaches the eight-cell stage does the increase in volume begin to mirror the increase in cell number. Because the zygote is an especially large cell, this lack of correspondence between an increase in cell number and growth may be unusual, but it points out the potential problem in equating an increase in cell number with growth. Although cell number may not always be a reliable measure of plant growth, under most circumstances dividing cells, particularly in meristems, double in volume during their cell cycle. Therefore, an increase in cell number, such as the increase brought about by the activity of the apical meristems, does contribute to plant growth. However, the largest component of plant growth is the rapid cell expansion that occurs in the subapical region after cell division ceases.

Because all the cells of the plant axis elongate under normal conditions, the greater the number of cells produced by the apical meristem, the longer the axis will be. For example, when Arabidopsis plants are transformed with a gene that encodes cyclin, a key component of the cell cycle regulatory machinery, the cells of the apical meristem progress through their cell cycles more rapidly, so more cells form per unit time.

As a result, the roots of these transgenic plants have more cells and are substantially longer than the roots of wild-type plants grown under similar conditions. New cells form continually in the apical meristems. With each new round of cell division and associated cell expansion, the older derivatives are displaced a small distance from the apex. As the cells recede farther from the apex, the rate of displacement is greatly accelerated. By viewing plant growth as a process of cell displacement from the apex, we can apply the principles of kinematics.

RESEARCH ON PLANT DEVELOPMENT

Plants and animals are separated by about 1.5 billion years of evolutionary history. They have evolved their multicellular organisation independently but using the same initial tool kit—the set of genes inherited from their common unicellular eucaryotic ancestor. Most of the contrasts in their developmental strategies spring from two basic qualities of plants. First, they get their energy from sunlight, not by ingesting other organisms. This dictates a body plan different from that of animals. Second, their cells are enclosed in semi rigid cell walls and cemented together, preventing them from moving as animal cells do. This dictates a different set of mechanisms for shaping the body and different developmental processes to cope with a changeable environment. Animal development is largely unaffected by environmental changes, and the embryo generates the same genetically determined body structure unaffected by external conditions.

The environment, by contrast, dramatically influences the development of most plants. Because they cannot match themselves to their environment by moving from place to place, plants adapt instead by altering the course of their development. Their strategy is opportunistic. A given type of organ—a leaf, a flower, or a root, say—can be produced from the fertilized egg by many different paths according to environmental cues.

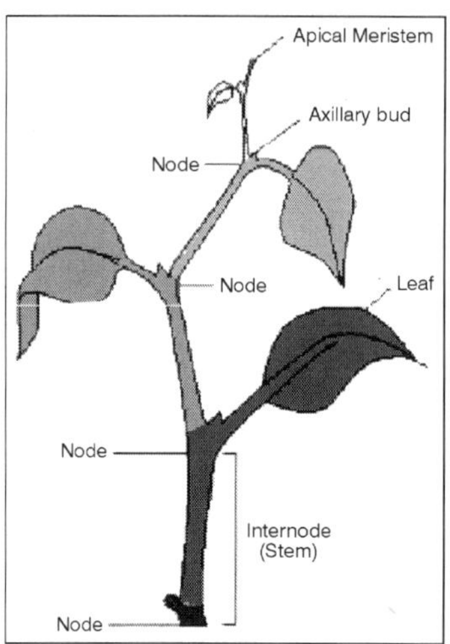

Fig. An Example of the Modular Construction of Plants. Consists of a Stem, a leaf, and a Bud Containing a Potential growth centre, or *Meristem.* The Bud forms at the Branch Point, or *Node,* where the leaf Diverges from the Stem. Modules arise Sequentially from the Continuous Activity of the Apical Meristem.

A begonia leaf pegged to the ground may sprout a root; the root may throw up a shoot; the shoot, given sunlight, may grow leaves and flowers. The mature plant is typically made of many copies of a small set of standardised modules, as described in (Figure-1). The positions and times at which those modules are generated are strongly influenced by the environment, causing the overall structure of the plant to vary. The choices between alternative modules and their organisation into a whole plant depend on external cues and long-range hormonal signals that play a much smaller part in the control of animal development. But although the global structure of a plant—its pattern of roots or branches, its numbers of leaves or flowers—can be highly variable, its detailed organisation on a small scale is not. A leaf, a flower, or indeed an early plant embryo, is as precisely specified as any organ of an animal, possessing a *determinate* structure, in contrast with the *indeterminate* pattern of branching and sprouting of the plant as a whole. The internal organisation of a plant module raises essentially the same problems in the genetic control of pattern formation as does animal development, and they are solved in analogous ways.

ARABIDOPSIS AS A MODEL FOR PLANT

Flowering plants, despite their amasing variety, are of relatively recent origin. The earliest known fossil examples are 125 million years old, as against 350 million years or more for vertebrate animals. Underlying the diversity of form, therefore, there is a high degree of similarity in molecular mechanisms. As we shall see, a small genetic change can transform a plant's large-scale structure; and just as plant physiology allows survival in many different environments, so also it allows survival of many differently structured forms. A mutation that gives an animal two heads is generally lethal; one that doubles the number of flowers or branches on a plant is generally not. To identify the genes that govern plant development and to discover how they function, plant biologists have selected a small weed, the common wall cress *Arabidopsis thaliana,* as their primary model organism. Like *Drosophila's* or *Caenorhabditis elegans,* it is small, quick to reproduce, and convenient for genetics. It can be grown indoors in Petri dishes or tiny plant pots in large numbers and produces hundreds of seeds per plant after 8 to 10 weeks. It has, in common with *C. elegans,* a significant advantage over *Drosophila* or vertebrate animals for genetics: like many flowering plants, it can reproduce as a hermaphrodite because a single flower produces both eggs and the male gametes that can fertilize them. Therefore, when a flower that is heterozygous for a recessive lethal mutation is self-fertilized, one-fourth of its seeds will display the homozygous embryonic phenotype. This makes it easy to perform genetic screens Figure and so to obtain a catalogue of the genes required for specific developmental processes.

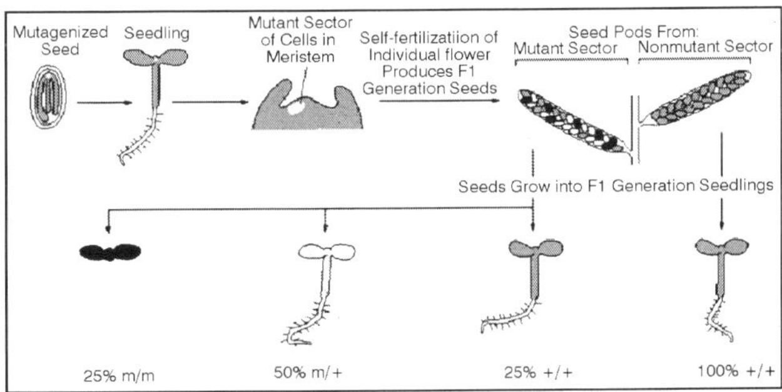

Fig. Arabidopsis Thaliana. This Small plant is a Member of the Mustard (or Crucifer) Family. It is a weed of no Economic use but of Great Value for Genetic Studies of Plant Development

Fig. Production of Mutants in a Rabidopsis.

Genes Control the Development of Arabidopsis

Arabidopsis has one of the smallest plant genomes—125-million nucleotide pairs, on a par with *C. elegans* and *Drosophila*—and the complete DNA sequence is now known. It contains approximately 26,000 genes. Cell culture and genetic transformation methods have been established, as well as vast libraries of seeds carrying mutations produced by random insertions of mobile genetic elements, so that plants with mutations in any chosen gene can be obtained to order. Powerful tools are thus available to analyse gene functions. Although only a small fraction of the total gene set has been characterised experimentally as yet, functions can be tentatively assigned to many genes—about 18,000—on the basis of their sequence similarities to well-characterised genes in *Arabidopsis* and other organisms.

Even more than the genomes of multicellular animals, the *Arabidopsis* genome is rich in genes that code for gene regulatory proteins (Table). Some major families of animal gene regulatory proteins are greatly expanded, while others seem to be entirely absent, and there are large families of gene regulatory proteins in the plant that have no animal homologs.

Where homologous gene regulatory proteins (such as homeodomain proteins) can be recognised in both plants and animals, they have little in common with regard to the genes they regulate or the types of developmental decisions that they control, and there is very little conservation of protein sequence outside the DNA-binding domains. *Arabidopsis* is like multicellular animals in possessing many genes for cell communication and signal transduction (1,900 genes out of 18,000 classified).

The Wnt, Hedgehog, Notch, and TGFβ signaling mechanisms are all absent in *Arabidopsis*. In compensation, other signaling pathways peculiar to plants are highly developed. Cell-surface receptors of the tyrosine kinase class seem to be entirely absent, although many of the signaling components downstream of these receptors in animals are present. Conversely, receptors of the serine/threonine kinase class are very plentiful, but they do not act through the same system of intracellular messengers as the receptor serine/threonine kinases in animals. Substantial sets of genes are devoted to developmental processes of special importance in plants: more than 700 for synthesis and remodelling of the plant cell wall, for example, and more than 100 for detecting and responding to light.

TRANSITION AND VEGETATIVE OF PLANT REPRODUCTIVE DEVELOPMENT

The rotation of our planet results in regular changes in environmental cues such as day length and temperature, and organisms have evolved a molecular oscillator that allows them to anticipate these changes and adapt their development accordingly. In many plants, the transition from vegetative to reproductive growth is controlled by photoperiod, which synchronizes flowering with favourable seasons of the year. One of the most important developmental decisions in a plant's life is when to switch from vegetative to floral (reproductive) growth. If flowering is initiated at the wrong time of the year, it will affect the number of seeds produced and significantly reduce reproductive success. In addition to its role in directly initiating flowering, light also serves as an entraining signal in resetting a plant's circadian clock. It is likely that all five phytochromes regulate red/far-red entrainment signals. The circadian system is crucial for photoperiod perception, as alterations in clock function alter the induction of flowering time.

In fact, many genes that regulate clock function were originally isolated as floral-timing mutants. From work on toc1, cca1 and lhy mutant lines, a molecular model for the central oscillator was proposed. In this model, TOC1 expression in the night activates CCA1 and LHY; the resulting CCA1/LHY expression in the morning represses TOC1 expression.

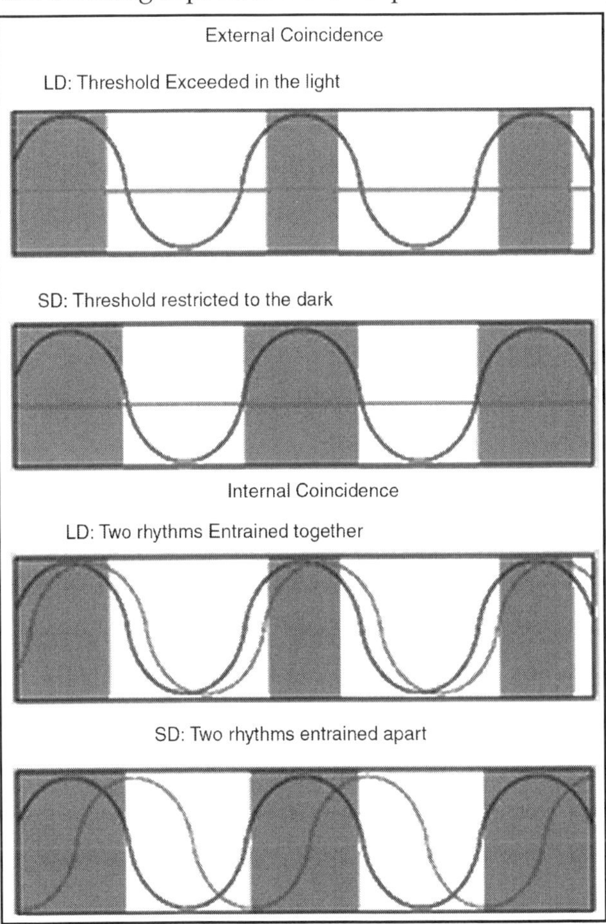

Fig. Two Models for Photoperiodic Timing

This repression is relieved by decay in CCA1/LHY expression, so that TOC1 levels increase again in the subsequent night, closing the 24 hour loop. New data argue against this model, however, as it appears that LHY and CCA1 are not absolutely required for circadian function. This has led to an alternative model, in which sequential periodic expression of TOC1 and its four homologs generates the oscillator.

Regardless of the exact molecular details of the oscillator, it is clear that the circadian system integrates photoperiod Perception and the UV/blue-light-absorbing cryptochromes. Mutations in the red-light receptor phytochrome B result in early flowering, particularly in the absence of an inductive photoperiod. This establishes this receptor as a repressor of flowering time. Mutations in cryptochrome 2 cause delayed flowering time under inductive photoperiods.

Thus, cryptochrome 2 is a promotive receptor of flowering time. The competition between these two receptors could act in the determina-tion of repression or activation of reproductive development. Molecular genetic analyses have identified photoreceptors and light-signalling components, and components of the circadian system, which are essential for a plant to make a proper photoperiodic response. Independent studies by the Carre and Kay groups have now provided a molecular foundation for understanding how light perception is integrated with the circadian system in the generation of a photoperiodic response. Their work supports the classical 'external coincidence' model which functions through the circadian expression of the CONSTANS gene and light activation of the encoded protein.

In the external-coincidence model, a physiological response, such as flowering, is triggered when light perception coincides with the time when expression of a circadian-regulated gene exceeds a required threshold. The long days (LD), light is perceived both at dawn and dusk of the expression phase, whereas under short days (SD), threshold expression is restricted to the dark. In the internal-coincidence model the effect of light is simply to entrain two distinct circadian oscillators. In the example shown, long days cause the two rhythms to be entrained with similar phases; this could generate two regulatory molecules that require each other's activity for physiological function. Under short days, the phases of the two entrained rhythms are further apart; this could restrict the simultaneous expression of two factors, thus inhibiting their co-action.

The link in Arabidopsis between light perception, the oscillator and flowering time appears to be the transcription factor CONSTANS (CO). Loss-of-function co mutants flower late in inductive long days, whereas ectopic overexpression of CO promotes early flowering independent of daylength. Further, CO expression level is reduced in late-flowering gi and lhy mutants, and elevated in early-flowering elf3 and elf4 mutants. Defects in CO expression can thus explain the opposite effects on flowering time of these mutations.

CO expression in Arabidopsis is modulated by light perception and the circadian oscillator, and so is dynamically regulated by daylength. In long-day photoperiods, CO abundance is high at the beginning and end of the photoperiod, but in short days, CO is restricted to the dark phase of the day. It was thus proposed that CO expression induces flowering as a function of daylength via a light-dependent post-transcriptional process that requires a threshold level of circadian transcribed CO.

Further support for this view comes from the finding that the expression of the CO target gene FLOWERING LOCUS T (FT) is restricted to when the expression of CO is coincident with light. Although this supports an external-coincidence model for flowering time, only the requirement of light was directly examined in these experiments When to flower is therefore a critical decision, and consequently multiple mechanisms have evolved to align flowering with optimal environmental conditions. But how does a plant recognize the presence of favourable conditions and integrate this information with its own endogenous developmental programme?

A new clue to this problem comes from the work of He *et al.* 2003. To dissect the molecular processes that initiate flowering and trigger the change from vegetative to reproductive growth, biologists have carried out intensive genetic studies of flowering time in the model plant Arabidopsi. This has led to the discovery of many genes involved in the regulation of flowering time and the development of a number of genetic models.

GENETIC PATHWAYS

Despite progress in identifying the genetic pathways involved, the mechanisms by which these flowering gene products modulate the floral transition is largely unknown. He *et al.* in 2003 elucidated one mechanism by which plants regulate flowering time. They have identfied a protein called FLOWERING LOCUS D (FLD), which removes acetyl groups from (deacetylates) histone proteins in chromatin containing the FLOWERING LOCUS C (FLC) gene. The FLC protein encoded by this gene is a member of the MADS-domain family of transcription regulators and is a strong repressor of flowering.

By deacetylating histones in FLC chromatin, FLD prevents transcription of FLC enabling the plant to flower. To ensure that flowering occurs at the correct time in Arabidopsis, floral initiation is regulated by the integration of signals from developmental pathways and environmental cues. These include pathways that monitor plant hormones, detect light quality and duration, and respond to prolonged exposure to cold (vernalization). A number of these pathways converge on the common target FLC. The autonomous promotion pathway regulates flowering independently of photoperiod (hence its name) by repression of FLC expression. FLD is one of six identified members of the autonomous pathway. Like other mutants in this pathway, fld mutants exhibit delayed flowering due to an increase in production of the flowering repressor FLC. In their new work, He and colleagues show that FLD is homologous to the human protein KIAA060, a component of the human Histone Deacetylase 1,2 (HDAC 1/2) complex. HDAC complexes remodel chromatin by removing acetyl groups from lysine residues in the tails of histones.

Hyperacetylated histones are associated with transcriptionally active genes, and hypoacetylated histones with transcriptionally silent chromosomal regions. To examine whether FLD may be a component of a plant HDAC complex, He and co-workers studied the acetylation state of histone H4 at the FLC locus. They immunoprecipitated specific chromatin fractions with an antibody against acetylated H4 histone tails and inspected the different fractions using polymerase chain reaction—a technique called chromatin immunoprecipitation (ChIP). They found thatin fld mutants of Arabidopsis, histone H4 tails in a specific region of the FLC locus are hyperacetylated compared with those in wild-type plants.

So, when the deacetylating activity of FLD is disrupted, FLC remains acetylated and is actively transcribed, resulting in a delay in flowering. HDACs are often recruited to their targets through interactions with cis-regulatory elements. To define these cis-elements, He and co-workers generated a series of internal deletions of the FLC gene and introduced them into plants lacking functional FLC. Removal of a 294–base pair (bp) region within intron 1 of FLC promoted hyperacetylation of FLC chromatin, FLC expression, and late flowering. Thus, deletion of this specific region of intron 1 has effects on FLC expression similar to those observed in fld mutant plants. Taken together these results show that FLD regulates FLC by deacetylating histones in FLC chromatin, providing evidence for chromatin regulation of FLC expression. The other components of the autonomous pathway also down-regulate FLC expression.

But He *et al.* demonstrate that, with the exception of FVE, they do not cause similar acetylation changes at the FLC locus. The autonomous pathway, therefore, must exploit multiple mechanisms to regulate FLC. Other components of the autonomous pathway include LD, which encodes a homeodomain protein; FPA and FCA, which encode RNA-binding proteins; and FY, which encodes a polyadenylation factor.

EXACT MECHANISM

However, the exact mechanism by which these components regulate FLC is still not known. Further studies examining how FLD regulation of FLC is integrated with information from other pathways modulating FLC will provide a useful model for understanding how regulatory networks converge on a single target. Although the homology of FLD to a component of the mammalian HDAC1/2 complex would imply that FLD is part of a similar complex in plants, this still needs to be demonstrated. Four genes with homology to HDAC1/2 have been found in the Arabidopsis genome. However, the effects of mutations in these genes do not resemble those in fld mutants. This may reflect redundancy among the HDACs such that no one mutation alters flowering time. Consistent with this, treatment with an antisense transgene that is likely to suppress several of the HDACs does result in late flowering.

The discovery of multiple FLD homologs in Arabidopsis raises the possibility that different FLD-like proteins are part of HDAC complexes with different target-site specificities. The He *et al.* study encourages further analysis of the presumptive FLD complex, which should provide a greater understanding of the evolutionary conservation of its constituent proteins and their involvement in transcriptional repression in both plants and mammals. Such an analysis may also reveal the ways in which plant development is more plastic and adaptable than animal development. Many plant cells are totipotent, so plants need versatile mechanisms to reprogram chromatin states. Many tools are now in hand to dissect these mechanisms and to address how they differ from those controlling reprogramming of animal genomes.

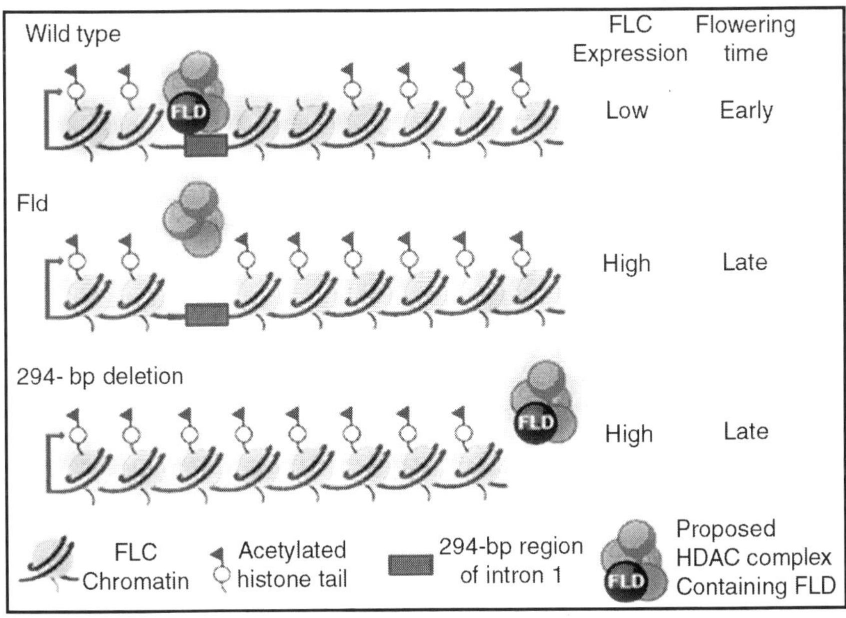

Fig. Chromatin Regulation and flowering. Regulation of the Floral Repressor FLC by FLD.

The acetylation states of FLC chromatin in wild-type plants (top), fld mutants (middle), and plants lacking a specific 294-bp region of FLC intron 1 (bottom). (Top) In wild-type plants, FLD acts as part of an HDAC complex, which interacts directly or indirectly with intron 1 of FLC to deacetylate specific regions of its chromatin. Consequently, FLC expression is reduced and the plant is able to flower. (Middle) In plants lacking functional FLD, FLD-dependent deacetylation of FLC is lost. Presumably, the HDAC complex containing mutant FLD can no longer be targeted to FLC. Therefore, the FLC locus remains acetylated, is actively transcribed, and flowering is delayed. (Bottom) Removal of a 294- bp region of FLC intron 1 also prevents FLD-dependent deacetylation of this locus possibly because the HDAC complex can no longer bind to FLC. Thus, FLC remains acetylated, and the gene is expressed, with a consequent delay in flowering.

In winter-annual types of Arabidopsis, flowering is delayed unless plants are vernalized. The delayed flowering is due to dominant alleles of FRIGIDA (FRI) and FLC. FRI elevates expression of the MADS-box transcriptional regulator FLC to levels that suppress flowering. Vernalization promotes flowering primarily by repressing FLC expression. The repressed state of FLC is maintained through mitotic cell divisions after a return to warm growing conditions. Many summer-annual accessions of Arabidopsis flower rapidly without vernalization because such accessions lack an active FRI allele or have a weak FLC allele and thus have low levels of FLC expression.

To ensure flowering in favourable conditions, many plants flower only after an extended period of cold, namely winter. In Arabidopsis, the acceleration of flowering by prolonged cold, a process called vernalization, involves downregulation of the protein FLC, which would otherwise prevent flowering. This lowered FLC expression is maintained through subsequent development by the activity of VERNALIZATION (VRN) genes. VRN1 encodes a DNA binding protein whereas VRN2 encodes a homologue of one of the Polycomb group proteins, which maintain the silencing of genes during animal development.

Here we show that vernalization causes changes in histone methylation in discrete domains within the FLC locus, increasing dimethylation of lysines 9 and 27 on histone H3. Such modifications identify silenced chromatin states in Drosophila and human cells. Dimethylation of H3 K27 was lost only in vrn2 mutants, but dimethylation of H3 K9 was absent from both VRN1 and VRN2, consistent with VRN1 functioning downstream of VRN2. The epigenetic memory of winter is thus mediated by a 'histone code' that specifies a silent chromatin state conserved between animals and plants.

5

Gene Transformation Methods in Plants

To achieve genetic transformation in plants, we need the construction of a vector (genetic vehicle) which transports the genes of interest, flanked by the necessary controlling sequences, *i.e.*, promoter and terminator, and deliver the genes into the host plant.

TRANSFORMATION OF PLANT GENOME

It is well known that transformation of plant genome using Agrobacterium tumefaciens is the exploitation of the process of pathogen infection. Normally, the initial response of plants to pathogen attacks is an oxidative burst with rapid and transient production of reactive oxygen species. This plant response indeed is a defence mechanism as ROS can kill the pathogenic bacteria or inhibit their growth. ROS production is usually followed by the hypersensitive response (HR) to pathogens leading to rapid cell death (necrosis).

Although the interaction between plant and Agrobacterium is not yet fully understood, several studies have reported necrosis and a poor survival rate of target plant tissues following Agrobacterium-mediated transformation. Many crop transformation methods that use NPTII selection have a common problem of regeneration of nontransgenic shoots while imposing kanamycin selection at the shoot formation stage. These nontransgenic shoots, referred to as "shoot escapes" that are different from "escapes" that we often refer as nontransgenic plants, occur in high frequencies in many crop species tested. For instance, the occurrence of ranging from 40% to 90% have been reported for apple, pear, banana, grapevine, citrange, sweet orange, and lime. In cauliflower Stipic et al. found that approximately 95% of shoots regenerated from selective media were shoot escapes. As with tissue necrosis, the exact reason for the occurrence of "shoot escapes" is unknown. A different but related issue is the inability of plant tissues to respond to culture manipulations for the desired outcome. Under normal tissue culture conditions, even without transformation, many plant species fail to respond to culture manipulations.

This phenomenon, referred to as "in vitro recalcitrance", has been ascribed to several factors including cellular incompetence and necrosis but, again, the exact reason(s) for this problem remain unclear.

The three common problems:

1. Browning/necrosis of transformed cells/tissues.
2. Shoot escapes in transgenesis.
3. In vitro recalcitrance, severely limit the number of transgenic plants that can be regenerated.

Although seemingly unrelated, recent research is beginning to unravel a common factor underlying these problems, the production of ROS, which can cause growth inhibition, cell death, or alter plant metabolic pathways leading to poor regeneration of plants and the production of shoot escapes. ROS include a number of chemically reactive molecules derived from oxygen whose functions have been reviewed.

Some of those molecules are extremely reactive, such as the hydroxyl radical, while some are less reactive (superoxide and hydrogen peroxide). Intracellular tree radicals, i.e., free, low molecular weight molecules with an unpaired electron, are often ROS and the two terms are, therefore, commonly used as equivalents. Free radicals and ROS can readily react with most biomolecules, starting a chain reaction of free radical formation. In order to stop this chain reaction, a newly formed radical must either react with another free radical or react with a free radical scavenger, such as an antioxidant. ROS are formed and degraded by all aerobic organisms, leading to either physiological concentrations required for normal cell function or excessive quantities, the state called oxidative stress.

Growing evidence has indicated that cellular reduction/oxidation (redox) status regulates various aspects of cellular function. Oxidative stress can elicit positive responses such as normal cellular proliferation, activation of transcription factors or gene expression, as well as negative responses such as growth inhibition or cell death. Excessive production of ROS often leads to oxidative stress, loss of cell function, and ultimately apoptosis or necrosis by its interference with various biomolecules, including proteins, lipids, and DNA. A balance between oxidant and antioxidant intracellular systems is hence vital for normal cell function, growth regulation, and adaptation to diverse growth conditions. Broadly, this review addresses some the biological aspects of ROS production and its manipulation to improve plant transformation. The emphasis will be on the causes and possible solutions to minimize regenerative recalcitrance due to cell death.

MAJOR PROBLEMS AND ANTIOXIDANTS

Hyperhydricity. Hyperhydricity, a physiological disorder occurring in plant tissue cultures, is associated oxidative damage. Hyperhydricity results in a general decrease in concentrations of reduced and oxidized pyridine nucleotides, reflecting a reduction in metabolic activity. The activities of antioxidant enzymes, such as superoxide dismutase (SOD), catalase, ascorbate peroxidase, and glutathione reductase, were higher in hyperhydric leaves than in normal leaves, indicating that hyperhydricity was associated with oxidative stress. Measurements of chlorophyll fluorescence provided evidence of oxidative damage to the photosynthetic machinery in the hyperhydric leaves because the photochemical efficiency of photosystem II, the effective quantum efficiency, and photochemical quenching were all lower in the hyperhydric leaves. It is reported that the inclusion of glutathione in a culture medium could suppress the hyperhydricity of calli, thereby promoting regeneration of plantlets in a desiccation-tolerant plant, Ramonda myconi.

In vitro recalcitrance of plants has been associated with ROS production. Higher levels of free radical activity were found in recalcitrant genotypes of potato and grape as well as in nonembryogenic calli (rice) and reduced embryogenic capacity calli (carrot) compared to responsive genotypes and embryogenic calli, respectively, The production of the free radicals might lead to an increase in lipid peroxidation and subsequently have a negative effect on the morphogenetic capacity of plant tissue cultures. Production of H_2O_2 coincided with the emergence of meristemoids and the formation of bud primordia in strawberry morphogenic calli. High O^-_2 level, low H_2O_2level, and little or no SOD activity were detected in calli possessing low organogenesis capacity.

Adding N,N- diethyldithiocarbonate, an SOD inhibitor, to the regeneration medium promoted O^-_2 production, inhibited H_2O_2 production and decreased the regeneration percentage, whereas the exogenous addition of H_2O_2 slightly promoted the potential for regeneration of shoot buds. The results suggest that H_2O_2 is directly correlated with the morphogenetic process in strawberry callus. Types II and III strawberry calli, which showed higher regeneration capacity, had five- and ninefold higher contents of intracellular H_2O_2, and three- and fourfold higher contents of intracellular O^-_2 compared to type I callus, which exhibited a much lower regeneration capacity. Types II and III calli also showed much higher activities of antioxidant enzymes than type I calli. The results indicated that ROS might play a dual role in the regeneration of strawberry calli.

On one hand, a certain level of a typical ROS may have a positive effect on strawberry regeneration. On the other hand, high levels of another ROS are inhibitory to the expression of totipotency in strawberry calli. Similar study found that H_2O_2 promoted somatic embryogenesis of gladiolus at 100 [mu]M but it inhibited shoot organogenesis. Antioxidants and the control of tissue browning and necrosis in plant tissue culture. Numerous studies reported tissue browning/necrosis leading to poor plant regeneration in vitro and successful use of antioxidants solving the problems in the tissue culture history. Most of the studies were associated with dicotyledonary plant species;

however, several antioxidants such as ascorbate, cysteine, dithiothreitol (DTT), glutathione, and tocopherol were also successfully used in monocotyledonary plant species. Among the studies, four of them had been well documented. Ziv and Halevy reported that using an antioxidant, DTT, controlled oxidative browning of tissues during in vitro propagation of Strelitzia reginae. Terminal and axillary buds, which were treated with a solution containing 0.04% DTT by submersing the buds for 24 h and then culturing the buds on an agar medium with charcoal or by culturing the buds on paper bridges in a liquid medium with 0.04% DTT, were able to develop to shoots. Also the degree of oxidative browning of the shoot tip expiants was reduced from a rating of 4 (without DTT) to a rating of 1.5. Shoot tips of apple (Malus pumild) rootstock M.26 turned brown immediately after being excised from expiants during in vitro propagation.

The browned shoot tips would neither proliferate nor develop into plantlets. Glutathione (GSH) was applied before the propagation to prevent browning of the shoot tip explants. Shoot development from the explants, which were treated with 0.1 mM GSH solution by dipping them in the solution prior to culture, was compared with that of the explants without GSH treatment (control). In the GSH treatment, 100% of shoot tips developed into normal shoots after 120 d, whereas the result from the control was 40%.

The results showed that application of GSH prior to the culture promoted the normal development of shoot tips. Apparently the major effect of GSH on the shoot tip development was the protection from browning of the shoot tips. For in vitro propagation of Protea cynaroides, oxidative browning of shoot segment expiants is a major problem. However, almost all expiants, which were immersed in a solution containing 100 mg/l ascorbic acid and 1,500 mg/l citric acid for 1 h, and grown under 16 h photoperiod, had a 100% of the expiants survived and developed shoots, while only 20% of the expiants survived without the antioxidant treatment.

Adonis amurensis is a perennial ornamental plant whose shoot tip expiant darkening is a major obstacle to establish in vitro propagation. Normally, about 20% of the expiants survive the initial stages of the culture. However, when they were soaked in an antioxidant solution containing 300 mg/l ascorbic acid and 300 mg/l citric acid for 30 min prior to the culture, survival rate was about two times higher (53.3%) than the nontreated control (23.0%).

Perl et al. found that short exposures of embryogenic calli of Vitis vinifera cv. Superior seedless to a diluted solution of Agrobacterium resulted in tissue necrosis. The necrosis seemed to be oxygen-dependent and correlated with elevated levels of peroxides. Therefore, the inclusion of both 1% PVPP and 2 mg/l DTT in coculture medium was found to improve expiant viability during and after cocultivation. They observed that the necrosis of the embryogenic calli was completely inhibited by these antioxidants while Agrobacterium virulence was not affected. These antioxidants enabled the recovery of stable transgenic grape plants resistant to hygromycin. In another study successfully used 1% PVPP and 2 mg/l DTT to control browning and necrosis in grape leaf tissue expiants in a coculture medium during Agrobacterium-mediated transformation.

It was found that the coculture of sugarcane leaf spindle sections with A. tumefaciens induced a rapid necrosis of the tissue. To minimize the necrosis, the leaf spindle expiants were incubated in a liquid medium containing 15 mg/1 ascorbic acid, 40 mg/1 cysteine, and 2 mg/1 silver nitrate for 60 h in the dark prior to inoculation with Agrobacterium and then cocultured in a cocultivation medium with the same antioxidants at the same concentrations, respectively, after the inoculation with Agrobacterium. By doing so, the percentage of the expiant viability was increased from 10% to 90%.

In addition, the percentage of GUS positive expiants was increased from 0% (without the antioxidants) to 100%. It is also investigated the effects of the three compounds on the necrosis of shoot meristem expiants prior to infection in rice transformation. The expiants, which were incubated in a liquid medium containing 20 mg/1 ascorbic acid, 40 mg/1 cysteine, and 5 mg/1 silver nitrate for 6 h in the dark, had an average of 6% of the each expiant area producing the necrosis, but the explants without the antioxidant treatment had 80% of the each expiant area producing the necrosis. The antioxidant treatment increased rice transformation efficiency from 17% without the antioxidant treatment to 30%.

Similarly, Olhoft et al. (2001a, b) increased Agrobacterium infection from 37% (without cysteine) to 91% in the soybean cotyledonary node region by including 400 mg/l cysteine in cocultivation medium, subsequently, resulting in a twofold increase in transformation efficiency (2.1% with cysteine vs. 0.9% without cysteine). Browning/necrosis on the expiant tissues was also reduced. The studies also showed that the frequency of transformed cells was increased only when cysteine was present during cocultivation of Agrobacterium and cotyledonary-node explants. Later, the authors reported the cocultivation medium with 3.3 mM cysteine, plus 1 mM DTT resulted a significant higher transformation efficiency (12.7%) than that either with cysteine alone (7.7%) or no cysteine (0.7%) when using hygromycin B selection.

Applying 400 mg/1 cysteine in the coculture medium increased both the frequency of transient a-glucuronidase (GUS) expression in target cells of com (56% with cysteine vs. 17% without cysteine) and the stable transformation frequency (6.2% with cysteine vs. 0.2% without cysteine). However, cysteine reduced the percentage of the immature zygotic embryos giving rise to embryogenic Type II callus from 99% when noninfected immature zygotic embryos incubated on cocultivation medium without cysteine to 52% when the embryos cultured on cocultivation medium containing cysteine. Zeng et al. further confirmed that the inclusion of 400 mg/l cysteine in a cocultivation medium increased stable transformation from 0.2% (without cysteine) to 5.9%

in soybean Agrobacterium-mediated transformation. Adding glutathione to the selection medium reduced hyperhydricity of leaf expiants, increased leaf expiant viability, and increased the frequency of transformation from 13% (without glutathione) to 45% in Agrobacteriummediated transformation of a desiccation- tolerant plant, Craterostigma plantagineum. It is investigated the effects of antioxidants, including ascorbic acid, sodium selenite, DL-a-tocopherol, and glutathione in a cocultivation medium, on ROS production, antioxidant activity, and stable transformation efficiency during peanut Agrobacterium-mediated transformation.

They found that glutathione, tocopherol, and selenite not only eliminated the formation of H_2O_2 produced in wound tissue during preparation of leaf expiants and their cocultivation with A. tumefaciens, but also decreased malondialdehyde (MDA) formation and enhanced the activities of the antioxidant enzymes such as superoxide dismutase (SOD) and catalase (CAT). The inclusion of 100 mg/l glutathione, 50 mg/l tocopherol, and 20 mg/l selenite in the cocultivation medium increased the transformation frequencies from 3.9% (no antioxidant) to 14.6%, 10.3%, and 12.4%, respectively.

Lipoic acid (LA) is a sulfur-containing compound involved in several multienzyme complexes such as pyruvate dehydrogenase, ct- ketoglutarate dehydrogenase, branched-chain keto acid dehydrogenase, and glycine decarboxylase complex. In animals, free LA and dihydrolipoic acid are metabolic antioxidants that are able to scavenge most reactive oxygen species, to recycle other antioxidants such as vitamin C, glutathione, and vitamin B, and to increase the expression of genes involved in the regulation of normal growth and metabolism as well as redox regulation of gene transcription. Therefore, LA was investigated in Agrobacterium-mediated transformation across five different plant species, and it has significantly improved the transformation methods, even for recalcitrant genotypes.

Frequencies of independent transgenic plant events were increased in soybean from 0.6% to 3.6%, in potato from 3% to 19%, in tomato from 28% to 94%, and in wheat from 2.9% to 5.4%. The frequency of putative transgenic embryo was increased in cotton from 41% to 61%. The frequencies of shoot escapes were reduced in soybean from 92% to 72%, in potato from 50% to 16%, and in tomato from 91% to 53%, under the optimal conditions.

This study also demonstrated that an increase of the transformation frequency and reduction of escapes in tomato were accompanied by a twofold reduction in the severity of the browning/necrosis of Agrobacterium-toansfomied tissues, a twofold increase in the survivability of the transformed tissues, a fourfold increase in the percentage of transgenic shoots, and a threefold reduction of the percentage of shoot escapes when using LA under optimal conditions. The application of LA in plant transformation has dramatically solved the three common problems in plant transformation: browning/necrosis of the transformed tissues, recalcitrance, and shoot escapes, which severely limit the number of transgenic plants produced. Analysis of published information indicates that the antioxidants used in plant transformation can be classified into two groups based on their biological function. The first group, consisting of ascorbic acid, cysteine, DTT, lipoic acid, and PVPP, functions to reduce expiant necrosis, increase viability of expiants, and improve transformation efficiency while the other one, which include glutathione, selenite, and alpha-tocopherol, reduces hyperhydricity and ROS and increases transformation efficiency.

PROGRAMMED CEU IN PLANT

It is reported that the cell death triggered by Agrobacterium spp. in maize tissues had several features of apoptosis, which were DNA fragmentation, and cytological changes, and cytochrome c release. She further showed the two antiapoptotic genes from baculovirus,and lap had the ability to prevent the onset of the apoptosis in maize tissues. is reported to act as a direct inhibitor of caspases whereas lap may act upstream to prevent activation of caspases. Caspases control most events of apoptosis in vertebrates and in invertebrates and are responsible, either directly or indirectly, for the cleavage of cellular proteins.The proteins include nuclear proteins such as poly (ADP-ribose) polymerase, DNA- dependent protein kinase, and lamins, as well as actin. The evidence that these genes could affect the reaction of maize cells to an apoptotic stimulus strongly indicates that plant may have caspase-like proteases regulating apoptosis. A caspase- like proteolytic activity was also reported in tobacco plants undergoing an HR triggered by an infection with Tobacco mosaic virus. Molecular identification of ROS signaling during programmed cell death (PCD) in plant.

One of the first genes identified and isolated in ROS-associated plant cell death is LSDl that encodes a zinc-finger protein. This protein, together with two other zinc-finger proteins (LOLl and LOL2), could act as a molecular rheostat, sensing changes in ROS homeostasis, thereby controlling apoptosis through the regulation of apoptotic genes. The inactivation of the Arabidopsis Executerl gene completely abolished singlet oxygeninduced cell death. The chloroplastic protein, which is encoded by Executerl, might perceive nonscavenged singlet oxygen species within the chloroplast.

Other important stress signal transducers include mitogen-activated protein kinases (MAPKs) that act upstream of the oxidative burst during ozone treatment and the HR. The primary ROS activated tobacco MAPK is the salicylic acidinduced protein kinase, which is required during harpindependent PCD. A MAPKKK of alfalfe activates cell death induced by H^sub 2^O^sub 2^ through a specific MAPK-scaffolding action. A recent finding demonstrates the conserved Arabidopsis BCL2associated athanogene protein has been shown to be induced by H^sub 2^O^sub 2^ and capable of provoking PCD in both yeast and plants.

Mechanism to control programmed cell death associated with ROS in plant. Several recent studies have begun to unravel the possible regulatory mechanisms of programmed cell death associated with ROS in plants. The most extensively studied form of plant PCD is HR to pathogen infection. Recent evidence indicates that autophagy, which is induced during the plant defence response, is one mechanism by which HR-PCD is controlled.

Autophagy is a process hi which cytosol and organelles are sequestered within doublemembrane vesicles that deliver the contents to the lysosome/ vacuole for degradation and recycling of the resulting macromolecules. Several AuTophaGy (ATG) genes, which are required for autophagic activity, have identified via genetic screens in yeast. The findings of Liu et al. imply that there is a prodeath signal that moves out of the HR lesion into the surrounding uninfected tissue. Therefore, autophagy most likely alters the induction, movement and/or recognition of the pro-PCD signal.

Autophagy is not required for HR-PCD execution but is required to limit PCD to the infection sites in plant and might prevent a prodeath signal from initiating PCD in healthy tissue. Another study also suggests that autophagy has an anti-PCD function in innate immunity. However, examination of the function of ATG genes in other higher eukaryotes indicates that autophagy might have a dual role: pro- and/or anti-cell death. The study from Torres and Dangl suggest that ROS and nitric oxide are possible candidates for a pro-PCD signal. As extracellular ROS are produced before the onset of HR-PCD, ROS have been thought to function as pro- PCD signals, either by directly killing the pathogen or by acting as signaling molecules that mediate defence responses.

Apoptosis functions as an important defence strategy by host cells against viral invasion. Many viruses contain the antiapoptotic genes to block the defence-by-death response of host cells. The expression of antiapoptotic genes such as iap and p35 from baculoviruses, ced-9 from Caenorhabditis elegans and bcl-2 from humans in tobacco, tomato, and passion fruit plant (Passiflora spp) suppressed the extensive cell death caused by fungal and bacterial pathogens and also enhanced resistance to some abiotic stresses such as wounding, salt, cold, UV, and herbicides. Particularly, baculovirus p35 and iap genes were expressed in com embryos and embryogenic calli and their expression reduced tissue browning 3 d after cocultivation with A. tumefaciens. These antiapoptotic genes can be engineered to be expressed in plants for controlling tissue browning/necrosis and enhancing plant transformation efficiency, especially in recalcitrant or poorly transformable plant species.

POTENTIAL MECHANISMS OF ANTIOXIDANTS

Exposure of plant tissues to Agrobacterium during plant transformation leads to browning/necrosis of targeted cells/tissues, which affect transformation efficiency. Browning/necrosis of targeted cells/tissues affects plant transformation in two ways. Browning/necrosis may occur in transformed cells within expiant tissues, inhibiting regeneration of the transformed cells/tissues. secondly, necrotic tissues are known to accumulate antimicrobial substances, which may inhibit the potential of Agrobacterium to colonize plant cells and transfer T-DNA. Thirdly, the active release of chemical signals, which induce the vir genes in Agrobacterium, occurs only in living cells.

This also could reduce the potential of Agrobacterium to transfer T-DNA into plants. The inhibition of regeneration of transformed cells/tissues may promote growth of non-transformed cells/tissues even under selective pressures and, subsequently, result in production of shoot escapes. With regard to Agrobaterium, the ROS, produced during attempted transfection could be toxic enough to directly kill the attacking Agrobacterium, thereby subsequently preventing Agrobacterium from colonizing plant cells and transferring T-DNA into plants.

It is observed that elevated levels of peroxidase activity in grape tissues correlated with Agrobacterium-induced necrosis in the host during Agrobacterium-mediated transformation. Peroxidase is known to mediate oxidative cross-linking of structural proteins in the cell wall and the Agrobacterium-induced increase in peroxidase activity in grape tissues could further confirm the role of oxidative burst in HR in plants to Agrobacterium infection. It is demonstrated that at least two genes residing within the T-DNA region of Agrobacterium are responsible for inducing necrosis in grape tissues.

Furthermore, the aviR gene in Agrobacterium vitis was found to be associated with Agrobacterium-induced HR. aviR is homologous to luxR, which implies that the ^groeactentww- induced HR is regulated by a quorum-sensing mechanism. The Agrobacterium-mdaced HR could lead to rapid and large generation of ROS in target plant cells, resulting in plant cells/tissues necrosis, oxidative stress on the invading Agrobacterium cells, production of toxic antibacterial substances, and the deleterious effects on DNA molecules, especially at the site of oxidative burst. All of these factors could significantly reduce the efficiency of stable transformation of plants.

Preliminary research on Agrobaterium-mduced plant cell death has shown that Agrobacterium likely causes browning/necrosis of transformed plant cells/tissues in vitro and ROS production during Agrobacterium infection in vivo induces necrosis. The ROS could kill the attacking Agrobacterium, thereby preventing Agrobacterium from infecting plant cells/tissues and delivering T-DNA into plants. Also, the necrosis prevents regeneration of transformed cells/tissues. Using antioxidants in plant tissue culture and transformation reduces the browning/necrosis of nontransformed and transformed cells/tissues and the frequencies of shoot escapes. In addition, the antioxidants increase the stable transformation efficiencies across dicotyledonary and monocotyledonay plant species, indicating their potential roles of controlling ROS for efficient production of transgenic plants. Advancing Agrobacterium-mediated transformation technology requires an understanding of the mechanisms of Agrobacterium- induced browning/ necrosis and the roles of antioxidants in facilitating plant transformation. The functions of the antioxidants described above have raised important questions. Do ROS produce when Agrobacterium infection takes place during in vitro transformation? Do the ROS cause the browning/necrosis of transformed cells/ tissues?

Do the antioxidants scavenge and/or down-regulate the ROS through regulating the expression of genes involving ROS-generating systems such as NADPH oxidase or pH-dependent cell wall peroxidase? Do the ROS scavenging and/or down-regulating activities prevent the browning/necrosis of transformed cells/tissues? Or do the antioxidants alter the expression of genes that play an important role in the regeneration of transformed cells/ tissues by regulating the production of ROS that damage cells/tissues, or by activating other metabolic pathways? Once these questions are addressed, we can control apoptotic responses during wounding and Agrobacterium infection in plant transformation more effectively. This will undoubtedly lead to the development of more efficient plant transformation systems.

PROCESS OF GENE TRANSFER IN PLANTS

The process of gene transfer from *Agrobacterium tumefaciens* to plant cells implies several essential steps:

- Bacterial colonisation
- Induction of bacterial virulence system,
- Generation of T-DNA transfer complex
- T-DNA transfer and
- Integration of T-DNA into plant genome.

Supported by the most recent experimental data and accepted hypothesis on T-DNA transfer we presented a hypothetical model depicting the most important stages of this process.

Bacterial Colonisation

Bacterial colonisation is an essential and the earliest step in tumour induction and it takes place when *A. tumefaciens* is attached to the plant cell surfac. Mutagenesis studies show that non-attaching mutants loss the tumour-inducing capacity.

The polysaccharides of the *A. tumefaciens* (lipopolysaccharides (LPS), and capsular polysacharides (K-antigens)) are proposed to play an important role in the colonising process. The LPS are integral part of outer membrane and capsular polysaccharides (K-antigens), lacking of lipid anchor, have strong anionic nature. There are evidences that capsular polysaccharides may play specific role during the interaction with the host plant. The chromosomal 20kb *att* locus contains the genes required for successful bacterium attachment to the plant cell. This locus has been extensively studied. Insertions in the left 10 kb side of this region produce avirulent mutants that could restore its attachment capacity if the culture medium is previously conditioned by the incubation of wild-type virulent bacterium with plant cells.

In contrast, mutational insertion in the 10 kb right side of the *att* locus results in the irreversible loss of attachment capacity, which can not be restored by conditioned medium. These results suggest that genes, placed at *att* left side, are involved in molecular signalling events, while the right side genes are likely to be responsible for the synthesis of fundamental components.

Induction of Bacterial Virulence System

The T-DNA transfer is mediated by products encoded by the 30-40 kb *vir* region of the Ti plasmid. This region is composed by at least six essential operons (*vir A, vir B, vir C, vir D, vir E, virG*) and two non-essential (*virF, virH*). The number of genes per operon differs, *virA, virG* and *virF* have only one gene; *virE, virC, virH* have two genes while *virD* and *virB* have four and eleven genes respectively. The only constitutive expressed operons are *virA* and *virG*, coding for a two-component (VirA-VirG) system activating the transcription of the other *vir* genes, and has structural and functional similarities to other already described for other cellular mechanisms.

VirA is a transmembrane dimeric sensor protein that detects signal molecules, released from wounded plants. The signals for VirA activation include acidic pH, phenolic compounds, such as acetosyringone, and certain class of monosaccharides which acts sinergically with phelolic compounds. VirA protein can be structurally defined into three domains: the periplasmic or input domain and two transmembrane domains (TM1 and TM2). The TM1 and TM2 domains act as a transmitter (signaling) and receiver (sensor). The periplasmic domain is important for monosaccharide detection. Within the periplasmic domain, adjacent to the TM2 domain is an amphipatic helix, with strong hydrophilic and hydrophobic region.

This structure is characteristic for other transmembrane sensor proteins and folds the protein to be simultaneously aligned with the inner membrane and anchored in the membrane. The TM2 is the kinase domain and plays a crucial role in the activation of VirA, phosphorilating itself on a conserved His-474 residue) in response to signalling molecules from wounded plant sites. Monosaccharide detection by VirA is important amplification system to respond to low levels of phenolic compounds. The induction of this system is only possible through the periplasmic sugar (glucose/galactose) binding protein ChvE, which interacts with VirA. Recent studies for determination of VirA regions, important for its sensing activity suggested the position, which may be involved on TM1-TM2 interaction. This interaction causes the exposure of the amphipathic helix to small phenolic compounds and suggests a putative model for the VirA-ChvE interaction.

Activated VirA has the capacity to transfer its phosphate to a conserved aspartate residue of the cytoplasmic DNA binding protein VirG. VirG functions as transcriptional factor regulating the expression of vir genes when it is phosphorilated by VirA. The C-terminal region is responsible for the DNA binding activity, while the N-terminal is the phosphorilation domain and shows homology with the VirA receiver (sensor) domain.

The activation of vir system also depends on external factors like temperature and pH. At temperatures greater than 32°C the vir genes are not expressed because of a conformational change in the folding of VirA induce the inactivation of its properties. The effect of temperature on VirA is suppressed by a mutant form of VirG ($VirG^c$), which activates the constitutive expression of the vir genes. However, this mutant can not confer the virulence capacity at that temperature to Agrobacterium, probably because the folding of other proteins actively participate in the T-DNA transfer process are also affected at high temperature.

Generation of T-DNA Transfer Complex

The activation of *vir* genes carries out the generation of single-stranded (ss) molecules representing the copy of the bottom T-DNA strand. Any DNA placed between T-DNA borders will be transferred to the plant cell, as single strand DNA, and integrated into plant genome. These are the only cis acting elements of the T-DNA transfer system. The proteins VirD1 and VirD2 play the key role in this step are, recognising the T-DNA border sequences and nicking (endonuclease activity) the bottom strand at each border. The nick sites are assumed as the initiation and termination sites for T-strand recovery. After endonucleotidic cleavage VirD2 remains covalently attached to the 5'-end of the ss-T-strand.

This association prevents the exonucleolitic attack to the 5'-end of the ss-T-strand and distinguishes the 5'-end as the leading end of the T-DNA transfer complex. VirD1 interacts with the region where the ss-T-strand will be originated. Experiments *"in vitro"* experiments evidenced that for cleavage of supercoiled stranded substrate by VirD2 is essential the presence of VirD1. The simultaneous restoration of the excised ss-T-strand is evolutionarily related to other bacterial conjugative DNA transfer processes, which include the generation of the single strand DNA.

Extensive mutation or deletion of the right T-DNA border is followed by almost completely loss of T-DNA transfer capacity, while at the left border resulted in lower transfer efficiency. This fact indicates that T-strand synthesis is in 5' to 3' direction, and it is initiated at the right border and that the termination process take place even when the left border is mutated or completely absent, although at lower efficiency. The presence of an enhancer or "overdrive" sequence next to the right border which specifically recognised by VirC1 protein could make the difference between both T-DNA borders.

Two Models for the Translocation of T-DNA-Complex

The transferring vehicle to the plant nucleus is a ssT-DNA-protein complex. According to the most accepted model, the ssT-DNA-VirD2 complex is coated by the 69 kDa VirE2 protein, a single strand DNA binding protein. This co-operative association prevents the attack of nucleases and, in addition, extends the ssT-DNA strand reducing the complex diameter to approximately 2 nm, making easier the translocation through membrane channels. However, that association does not stabilises T-DNA complex inside *Agrobacterium.* VirE2 contains two plant nuclear location signals (NLS) and VirD2 one. This fact indicates that both proteins are presumably to play important role once the complex is in the plant cell mediating the complex uptake to the nucleus.

VirE1 is essential for the export of VirE2 to the plant cell. Bacterial strains mutated in *virE1*, cannot export VirE2 and resulted in its accumulation inside the bacterium. Such mutants can be complemented if coinfecting with strain that can export VirE2, indicating that this protein can be exported independently and that for transmission event does not necessary the transfer of VirE2 as part of the ss-T-DNA complex and that is possible the transfer of naked T-DNA to the plant cell. From these experimental evidences an alternative model was brought. This model proposes that the transfer complex is a single-strand DNA covalently bound at its 5'-end with VirD2, but uncoated by VirE2. The independent export of VirE2 to plant cell is presented as natural process, and once the naked ssT-DNA-VirD2 complex is inside the plant cell, it is coated by VirE2.

Previous researches described the role of 9.5 kb *virB* operon in the generation of a suitable cell surface structure for the ssT-DNA complex transfer from bacterium to plant. The VirD4 protein is also required for the ss-T-DNA transport. The function of VirD4 is the ATP-dependent linkage of protein complex necessary for T-DNA translocation. VirB are proteins that present the hydrophathy characteristics similar others membrane-associated proteins. VirD4 protein is a transmembrane protein but predominantly located at the cytoplasmatic side of the cytoplasmic membrane.

Comparative studies showed highly degree of homology between the *virB* operon and transfer regions of broad host range (BHR) plasmid in genetic organisation, nucleotide sequence and protein function. Both systems delivery non-self transmissible DNA-protein complex to recipient host cell. In addition, they have the capacity to DNA interkingdom delivery suggesting that T-DNA transfer apparatus and conjugation systems are related and probably evolved from a common ancestral. The majority of VirB proteins are assembled as a membrane-spanning protein channel involved both membranes. Except for

VirB11, they have multiple periplasmic domains. VirB1 is the only member of VirB proteins found in the extracellular milieu although it is possible that some of the other VirB proteins may be redistributed during the process of biogenesis and functioning of the transcellular conjugal channel. That could be the case of the VirB2, which is translated as a 12 kDa proprotein, and later is later proteolically processed to its mature 7 kDa functional form.

VirB4 and VirB11 are hydrophilic ATPases necessary for active DNA transfer. Vir B11 lacks continuous sequence of hydrophobic residues, motivating of periplasmic domains. These characteristics are atypical for this type of protein and evidence the possible dinamic co-existence of different conformational forms *"in vivo"*. VirB4 tightly associates with the cytoplasmic membrane. It contains two putative extracellular domains conferring transmembrane topology to this protein, which is presumed to allow the ATP-dependent conformational change in the conjugation channel. Probably, the functional forms of VirB4 and VirB11 are homo- and heterodimmers. The VirB7-VirB9 heterodimmer is assumed to stabilises other Vir proteins during assembly of functional transmembrane channel.

Recent studies identified some of the initial steps of biogenesis of ssT-DNA complex apparatus. Firstly, VirB7 and VirB9 monomers are exported to the membrane and processed. They interact each other to form covalently cross-linked homo- and heterodimmers. Although it is widely assumed the role of both types of dimmers in the biogenesis of the transfer apparatus, it is likely that only heterodimmers are the essential ones. Subsequently the VirB7-VirB9 heterodimmer is sorted to the outer membrane. The sorting mechanism has not been elucidated the next step implies the interaction with the other Vir proteins for assembling the transfer channel with the contribution of the transglycosidase.

Two accessory *vir* operons, present in the octopine Ti-plasmid are *virF* and *virH*. The *virF* operon encoding for a 23-kDa protein that functions once the T-DNA complex is inside the plant cells via the conjugal channel or independently, as it was assumed for VirE2 export. The role of VirF is seems to be related with the nuclear targeting of the ssT-DNA complex but its contribution is less important than in the case of VirF. The *virH* operon consists in two genes that code for VirH1 and VirH2 proteins. These Vir proteins are no essential but could enhance the transfer efficiency, detoxifying certain plant compounds that can affect the growth of bacterial. If that is the function of VirH proteins, they play a role in the host range specificity of bacterial strain for different plant species.

Integration of T-DNA into Plant Genome

Inside the plant cell, the ssT-DNA complex is targeted to the nucleus crossing the nuclear membrane. Two Vir proteins have been found to be important in this step: VirD2 and VirE2, which are the most important; and probably VirF, which has minor contribution to this process. The two NLS of VirE2 have been considered important for the continuos nuclear import of ss-T-DNA complex, probably by keeping both sides of nuclear pore simultaneously open. The nuclear import is probably also mediated by specific NLS-binding proteins, which are present in plant cytoplasm.

The final step of T-DNA transfer is its integration into plant genome. It is considered that the integration occurs by illegitimate recombination. According to this model, paring of a few bases provides just a minimum specificity for the recombination process by positioning VirD2 for the ligation. The 3′-end or adjacent sequences of T-DNA find some low homologies with plant DNA resulting in the first contact (synapses) between the T-strand and plant DNA forms a gap in 3'-5' strand of plant DNA. Displaced plant DNA is subsequently cut at the 3'-end position of the gap by endonucleases, and the first nucleotide of the 5' attached to VirD2 pairs with a nucleotide in the top (5'-3') plant DNA strand. The 3' overhanging part of T-DNA together with displaced plant DNA are digested away, either by endonucleases or by 3'-5' exonucleases. Them, the 5' attached to VirD2 end and other 3'-end of T-strand (paired with plant DNA during since the first step of integration process) joins the nicks in the bottom plant DNA strand. Once the introduction of T-strand in the 3'-5' strand of the plant DNA is completed, a torsion followed by a nick into opposite plant DNA strand is produced. This situation activates the repair mechanism of the plant cell and the complementary strand is synthesised using the early inserted T-DNA strand as a template.

VirD2 has an active role in the precise integration on T-strand in the plant chromosome. The release of VirD2 protein may provide the energy containing in its phosphodiester bond, at the Tyr29 residue, with the first nucleotide of T-strand, providing the 5'-end of the T-strand for ligation to the plant DNA. This phosphodiester bound can serve as electrophilic substrate for nucleophilic 3'-OH from nicked plant DNA. When the mutant VirD2 protein is transferred attached to the T-strand, the integration process take place with the loss of nucleotides at the 5'-end of the T-strand.

METHOD OF PLANT TRANSFORMATION

The first plant transformed by *Agrobacterium tumefaciens* was tobacco. Since that crucial moment in the development of plant science, a great progress in understanding the *Agrobacterium*-mediated gene transfer to plant cells has been archived. However, *Agrobacterium tumefaciens* naturally infects only dicotyledonous plants and many economically important plants, including the cereals, remained accessible for genetic manipulation during long time. For these cases alternative direct transformation methods have been developed. However *Agrobacterium*-mediated transformation have remarkable advantages over direct transformation methods in reducing the copy number of the transgene, potentially leading to fewer problems with transgene cosuppresion and instability. In addition, It is a single-cell transformation system and avoids the obtainment of mosaic plants, which are more frequent when direct transformation is used.

The monocots have been considered to be outside the *Agrobacterium* host range and other gene-transfer methods were developed for these plants. To develop this methodologies for a monocot plant it is important to take in consideration the critical aspects in the *Agrobacterium tumefaciens*-plant interaction, the cellular and tissue culture methodologies developed for that specie. The suitable genetic materials (bacterial strains, binary vectors, reporter and marker genes, promoters) and molecular biology techniques available in the laboratory, are necessary for selection of DNA to be introduce. This DNA must be expressible in plant making possible the identification of transformed plants in selectable medium and using molecular biology techniques test and characterise the transformation events.

Transformation is currently used for genetic manipulation of more than 120 species of at least 35 families, including the most major economic crops, vegetables, ornamental, medicinal, fruit, tree and pasture plants transformation methods. The idea, that some species can not accept the integration of foreign DNA in its genome and lack the capacity to be transformed is unacceptable under the increasing number species that have been transformed. The optimisation of *Agrobacterium tumefaciens*-plant interaction is probably the most important aspect to be considered. It includes the integrity of bacterial strain its correct manipulation as warranty of the virulence machine integrity and the study of reaction in wounded plant tissue, which may develop necrotic process in the wounded tissue or affect the interaction and release compounds inducers or repressors of *Agrobacterium* virulence system. The type of explant is also important fact and it must be suitable for regeneration allowing the recovering of whole transgenic plants. The establishment of method for efficient regeneration for one particular species is crucial for its transformation.

The mentioned aspects are important to establish transformation procedure for any plant but particularly for those species categorised as recalcitrant. In this category have been included cereals, legumes and woody plants, which are very difficult to transform or remain untransformed. Many species originally considered in this category has been transformed in recent years. One of these species, sugarcane, has been transformed in our laboratory.

Concluding Remarks

Agrobacterium tumefaciens is more than the causative agent of crown gall disease affecting dicotyledonous plants. It is also firstly the natural instance for the introduction of foreign gene in plants allowing its genetic manipulation. Similarities have been found between T-DNA and conjugal transfer systems are evolutionally related and apparently evolved from a common ancestral. Although the gene transfer mechanisms remain largely unknown, great progress has been obtained in practical implementation of transformation protocols for both dicotyledonous and monocotyledonous plants.

Particularly important is the extension of this single-cell transformation methodology to monocotyledonous plants. This advance has biological and practical implications. Firstly, because of advances of *A. tumefaciens*-mediated gene transfer over the direct transformation methods, which where the only way for genetic manipulation of economically important crops as cereals and legumes. Second, it has been demonstrated that T-DNA is transferred to dicot and monocot plants by an identical molecular mechanism. This confirmation implies that any plant can potentially be transformed by this method if suitable transformation protocol is established.

The *Agrobacterium*-mediated transformation protocols differ from one plant specie to other and, within species, from one cultivar to other. In consequence, the optimisation of Agrobacterium-mediated transformation methodologies requires the considered of several factors that can be determined in the successful transformation of one species. Firstly, the optimisation of *Agrobacterium*-plant interaction on competent cells from different regenerable tissues. Second, the development of suitable method for regeneration from transformed cells.

Undoubtedly, the development of transformation procedures based on A. tumefaciens-mediated gene transfer for new economically important species are advisable and the results obtained in last years evidence a promising future.

VECTOR-MEDIATED OR INDIRECT GENE TRANSFER

Among the various vectors used in plant transformation, the Ti plasmid of Agrobacterium tumefaciens has been widely used. This bacteria is known as "natural genetic engineer" of plants because these bacteria have natural ability to transfer T-DNA of their plasmids into plant genome upon infection of cells at the wound site and cause an unorganised growth of a cell mass known as crown gall. Ti plasmids are used as gene vectors for delivering useful foreign genes into target plant cells and tissues. The foreign gene is cloned in the T-DNA region of Ti-plasmid in place of unwanted sequences.

To transform plants, leaf discs (in case of dicots) or embryogenic callus (in case of monocots) are collected and infected with Agrobacterium carrying recombinant disarmed Ti-plasmid vector. The infected tissue is then cultured (co-cultivation) on shoot regeneration medium for 2-3 days during which time the transfer of T-DNA along with foreign genes takes place. After this, the transformed tissues (leaf discs/calli) are transferred onto selection cum plant regeneration medium supplemented with usually lethal concentration of an antibiotic to selectively eliminate non-transformed tissues. After 3-5 weeks, the regenerated shoots are transferred to root-inducing medium, and after another 3-4 weeks, complete plants are transferred to soil following the hardening (acclimatisation) of regenerated plants. The molecular techniques like PCR and southern hybridisation are used to detect the presence of foreign genes in the transgenic plants.

SIGNIFICANCE OF PLANTS CELL EXPANSION

Another significant observation of this early rapid growth is insensitive to the inhibition of transcription (RNA synthesis) and translation (protein synthesis) - ex. Actinomycin-D at high concentration inhibits RNA synthesis completely and cycloheximide (CHI) inhibits cytosol protein synthesis. This clearly suggests that the early rapid growth requires neither the synthesis of RNA nor the synthesis of proteins. But this early effect of auxin is totally inhibited by colchicines at 0.5 to 1 per cent concentration. Colchicine, an alkaloid from the tubers of Colchicum autumnale brings about the inhibition of tubulin (TB) polymerisation into long microtubules, which act as the cytoskeleton for mechanical support in the cytoplasm. The colchicines effect on early growth further indicates the important role played by the assembly of microtubules in providing a mechanical structure and the force for all expansion.

Increased metabolic activity, particularly the degradation of osmotically inactive starch, etc., leads to the increase of osmotic pressure (they are related to each other) and a steep DPD gradient is created. This facilitates the inward diffusion of water; consequently turgour pressure builds up within the cells. This force acts on all sides of the cell wall and forces the cell to expand in length and breadth. However recent investigations have shown that the cell elongation takes place under negative pressure. The above features suggest that the process or the mechanism of cell elongation is complex and multiphased, but highly regulated. Various theories have been proposed from time to time and none of them could prove their claim because of one or the other drawbacks. Here we will restrict to a comprehensive account of cell elongation.

Early phase: The auxin's effect in the early phase on cell growth is rapid. This is insensitive to the inhibitors of transcription and translation but insensitive to colchicines and respiratory inhibitors like cyanide, DNP etc. From this it can be deduced that cell elongation does not require new RNA and protein synthesis, but it requires ATP and ATP dependent tubulin polymerisation into microtubular cytoskeleton. Auxin at the early stages activates and increases the rate of respiratory metabolism. With the result ATP pool builds up rapidly. Auxin also activities membrane bound nucleating centre of tubulin polymerisation. These nucleating centres, utilising ATP and tubulin monomers found in the cytoplasm, start polymerising and soon build up long microtubules. These, in turn, start orienting parallel to the long axis of the cell, which probably requires cytochalasin-B sensitive microfilaments. With the rapid increase in microtubules, oriented in longitudinal plane, considerable force is exerted on the cell boundaries. As the microtubular polymerisation is building up, ATP ase/H* gets activated, and H* ions are pumped out into the interspaces of cellulose fibrils acidic pH in turn activates cellulases enzymes, which starts loosening up the cell-wall cellulose fibres. This loosening up of cell wall concomitant with acting as a mechanical force-the cell expands in length. This process starts 4-5 minutes after application of auxins and the elongation continues rapidly for about 20-30 minutes.

The late phase starts around 30-40 minutes after auxin application. This is slower than the first phase, but growth is steady. This phase is sensitive to Actinomycin-D, CHI, Colchicine and respiratory inhibitors, indicting that this process requires metabolic energy, tubulin-polymerisation, RNA synthesis as well as protein synthesis. As tubulin pool gets exhausted in the first phase, it has to be replaced. That means it requires the synthesis of RNA and proteins. Apart from these proteins, the cytoplasm requires the synthesis of other proteins, for the increase in the mass of protoplasm. It is quite possible that the early activation of early translation may have significant effect on gene activation. Thus it takes some time for the second phase of growth to set in. The second phase provides all materials for sustained growth.

Growth Movements: The phenomenon of bending movement of plants in response to light stimulus is called Phototropism, the growth of roots towards soil is called Geotropism. The curvature growth due to the touch is named as Thigmotropism, etc. Such differential growth movements are due to differential concentration of auxin. The stem tips and root tips have different optima of their auxin requirement to show maximum growth.

The mechanism of these movements will be explained in the chapter on growth movements. Auxin brings about phototropic movement because differential accumulation of auxin at exposed and darker side. *Root initiation*: Vegetative propagation of stem cuttings is a normal practice in horticulture, floriculture and agriculture. Out of some stem cuttings which are put into soil, a few produce roots easily but others may or not produce roots. The roots that develop from any part of the plant body other than the radicle are called adventitious roots. The development of such roots from the stem or any other part is a phenomenon of dedifferentiation. Stem is quite different fro the root with respect to its structure and function.

Still some of the tissues found in stem undergo dedifferentiation and develop into roots. This dedifferentiation and growth of roots from the stem is initiated by auxins Indole acetic acid, IBA, NAA, and other hormones. When applied to stem cuttings they induce the formation of roots. IBA has been found to be very good in this regard.This indicates that the pre-existing m RNA protein complex is drawn towards protein synthesis: it shows increased rate of protein synthesis. Some of the proteins thus synthesised in turn activate differential nuclear gene activation. As a result of it, a cascade of events occurs. Generally there is an increase in the synthesis of all cellular proteins. At the same time there is an enhanced rate of synthesis of some specific proteins. Also some new proteins are synthesised. Among these, tubulin, a membrane bound fraction, is synthesised at a high rate. It is believed that this protein along with microfilaments gets assembled into microtubules which orient themselves in a specific way so as to facilitate the polarity fixation for cell division.

In response to auxins, some of the cells in the pericycle region get stimulated and the required macromolecules are synthesised by differential gene activation. Thus these cells endowed with these new potentialities divide and redivide, and soon they organise into root primordial. They in turn develop into new roots which penetrate through cortical cells. The time taken for this new root formation, in response to auxin supply is just 24-36 hours. This effect of auxin induced new root formation has been well exploited by horticulturists and others.

The following diagram shows how auxins interact with cytokinins in eliciting growth responses. *Apical dominance*: The active growth of terminal bud inhibits the development of lower axillary bunds. This effect is called apical dominance. Many plants like conifers and other exhibit the apical dominance so predominantly that the plant grows tall with few branches and appear conical in shape. But plants like banyan, mango, tamarind do not show that much of apical dominance, hence these plants spread outwardly. This phenomenon of apical dominance has been attributed to the differential concentration of auxin and nutrient supply to the axillary buds.

Abscission: With the onset of winter season the leaves of deciduous plants turn yellow and fall off. The falling of the leaves is due to the formation of abscission layers at the base of the petioles. This layer consists of thin walled but suberised cells. The middle wall gets disintegrated and the leaves fall down. The development of this layer is stimulated by a plant hormone called ABA. But the application of auxin to such leaves prevent the formation of abscission layer and thus prevent the falling off the leaves. This has a greater application particularly in preventing the premature falling of fruits like apple, orange, etc.

Parthenocarpy: Development of fruits without fertilization is referred to as parthenocarpy and such fruits are seedless. Normally plants do not produce parthenocarpic fruits. Normally plants do not produce parthenocarpic fruits. But the application of auxin to flowers induces parthenocarpic fruit setting and prevents premature falling of fruits as well. Plants with triploid and many polyploid genomes do not set fruits. If such plants are sprayed with auxins at the time of flowering, fruit setting is enhanced. The fruits thus developed are large, seedless, and sweeter. Seedless oranges, lime, watermelon, sapota, ramphal and seethaphal, if they are produced this way, they have good market at the commercial level.

Gibberllic Acid

Gibberllic Acid is another natural hormone found distributed in higher plants. This was first discovered in Japan, later its structural and functional details have been elucidated by the people in the west. So far 52 or more different kinds have been identified. Among them GA2 2GA1 are the most common ones found in all plants. GA is a steroid like molecule in its structure. Young leaves are the sites in which this hormone is mediated by phytochromes and then it is transported to the regions of growth.

The most significant effect of GA on the growth of plants is it effects internodal cell elongation profoundly than any other hormones. Even genetic dwarfs (eg. Dwarf Pisum sativum) grow as tall as the natural tall plants. It also brings about a dramatic effect called Bolting and Flowering. Long day plants (which require 12-14 hrs of day light treatment for flowering) under short day conditions produce a cluster of leaves on highly shortened stem and never produce flowers. When such plants are sprayed with GA, the stem starts elongation with enhanced growth of internodes. Simultaneously the terminal buds starts producing floral branches. This phenomenon is called bolting and flowering. However both bolting and flowering are separated by time and space. Besides these above said effects, GA is essential for the mobilisation of food reserves for the germinating cereal grains like wheat and maize. At the time of germination, the young embryonic axis synthesises this hormone, which is soon released into endosperm and reaches the aleurone layer.

It is this layer, GA activates differential gene expression. As a result, specific m.RNAs is transcribed and later specific proteins like amylase are synthesised on translation. The amylase thus synthesised, acts upon the starch found in the endosperm, and the glucose thus released is drawn into the embryonic tissue which is in dire need of energy source. Notwithstanding the above effects, GA also effects parthenocarpy, rooting, germination, etc. This hormone also interacts with other natural hormones like auxin and cytokinin and brings about differential growth.

CYTOKININS

Another class of growth regulating hormones has been discovered by Miller et al and they are called Cytokinins. They are the derivatives or nucleotides. 6-Furfuryl amino purines, Benzyl amino purines and such compounds act as cytokinins.

Cytokinins are generally synthesised in the root tips, and then they are translocated to other regions. The endosperm of coconut milk, Borassus and of other plants is a rich source of cytokinins. Cytokinin level in plants is further regulated by other hormones like Auxins and GAs. On the contrary cytokinins control the synthesis of other hormones like Abscisic acid, a growth inhibitor. Cytokinin *per se* has an important role in the cytokinesis. But along with auxins, it effects differentiation of plant organs from the callus. Cytokinin by itself inhibits root formation, but in the presence of auxins, it induces organ formation like roots and shoots. If the ratio between auxin and cytokinin is high, roots are differentiated, but the lower ratio induces the formation of shoots.

The intermediate ratio stimulates the development of these hormones in differentiation at the molecular level is very interesting by unfortunately it is least understood. Cytokinins are also found to break dormancy of winter buds as well seeds. The exact mechanism by means of which they over come the dormancy is not known, still it has been thought, that they effect through the suppression of IBA which is responsible for dormancy. Another interesting effect that the cytokinins produce is the prolongation of senescence (Ageing). This effect is often referred to as Richmond and Lang's effect.

VECTORLESS OR DIRECT GENE TRANSFER

In the direct gene transfer methods, the foreign gene of interest is delivered into the host plant cell without the help of a vector. The methods used for direct gene transfer in plants are: Chemical mediated gene transfer, *e.g.*, chemicals like polyethylene glycol (PEG) and dextran sulphate induce DNA uptake into plant protoplasts. Calcium phosphate is also used to transfer DNA into cultured cells. Microinjection where the DNA is directly injected into plant protoplasts or cells (specifically into the nucleus or cytoplasm) using fine tipped (0.5 - 1.0 micrometer diameter) glass needle or micropipette. This method of gene transfer is used to introduce DNA into large cells such as oocytes, eggs, and the cells of early embryo.

Electroporation involves a pulse of high voltage applied to protoplasts/cells/tissues to make transient (temporary) pores in the plasma membrane which facilitates the uptake of foreign DNA. The cells are placed in a solution containing DNA and subjected to electrical shocks to cause holes in the membranes. The foreign DNA fragments enter through the holes into the cytoplasm and then to nucleus.

Particle gun/Particle bombardment - In this method, the foreign DNA containing the genes to be transferred is coated onto the surface of minute gold or tungsten particles (1-3 micrometers) and bombarded onto the target tissue or cells using a particle gun (also called as gene gun/shot gun/microprojectile gun).The microprojectile bombardment method was initially named as biolistics by its inventor Sanford. Two types of plant tissue are commonly used for particle bombardment- Primary explants and the proliferating embryonic tissues.

- *Transformation:* This method is used for introducing foreign DNA into bacterial cells, *e.g.* E. Coli. The transformation frequency (the fraction of cell population that can be transferred) is very good in this method. *E.g.* the uptake of plasmid DNA by E. coli is carried out in ice cold $CaCl_2$ (0-50C) followed by heat shock treatment at 37-450C for about 90 sec. The transformation efficiency refers to the number of transformants per microgram of added DNA. The $CaCl_2$ breaks the cell wall at certain regions and binds the DNA to the cell surface.
- *Conjuction:* It is a natural microbial recombination process and is used as a method for gene transfer. In conjuction, two live bacteria come together and the single stranded DNA is transferred via cytoplasmic bridges from the donor bacteria to the recipient bacteria.

Liposome mediated gene transfer or Lipofection - Liposomes are circular lipid molecules with an aqueous interior that can carry nucleic acids. Liposomes encapsulate the DNA fragments and then adher to the cell membranes and fuse with them to transfer DNA fragments. Thus, the DNA enters the cell and then to the nucleus. Lipofection is a very efficient technique used to transfer genes in bacterial, animal and plant cells.

6

Plant Cells Model and its Parts

PLANT CELLS

Plant cells differ from animal cells in that they have three different structures known as cell wall, vacuoles and plastids. However, they lack centrioles and intermediate filaments, which are present in animal cells. Read more on plant cell structure and parts.

Cell Membrane

It permits the waste material to exit the cell and regulates the movement of materials in and out of the cell. It also acts as a barrier between the inside of the cell and the outside, so that the chemical conditions on both sides can be different.

Cell Wall

It provides rigid structural support to the cell from which structures, like, leaves and stems are produced. It allows the circulation and distribution of water, minerals and nutrients in and outside the cell. The wall also controls the presence of pathogen microbes and development of tissues in the cell.

Golgi Body

It stores, packages, and circulates the lipids and proteins made in the endoplasmic reticulum. It stores the proteins into packages called vesicles.

Rough Endoplasmic Reticulum

It synthesises and exports proteins and glycoproteins throughout the cell.

Lysosomes

These organelles digest food by breaking down larger molecules into smaller ones, using special proteins.

Cytoplasm

It helps in maintaining the cell shape. Along with this, it plays an important role in the internal movement of cell organelles, cell mobility and muscle fibre contraction. It also distributes oxygen and nutrients to different parts of the cell.

Nucleolus

It is the most prominent structure in the nucleus wherein ribosomes are made.

Vacuole

The plant cell contains a large, single vacuole (an enclosed compartment), which is used to store water, compounds and minerals that help in plant growth.

Ribosomes

These are packets of RNA (Ribonucleic Acid) and are called as protein builders or synthesisers of the cell.

Chloroplasts

They are the food producers of the cell, containing chlorophyll, the green pigment essential for photosynthesis. Their main function is to generate sugars and starches.

Nucleus

It's the most important part of the cell, comprising chromosomes, *i.e.*, structures made up of genetic information that helps in cell growth and reproduction. Read more on cell nucleus: structure and functions.

Nuclear Envelope

It's an enclosure that surrounds nucleus and its contents. Unlike cell membrane, which has pores and spaces for RNA and proteins to pass through, it keeps the chromatin and nucleolus inside the nucleus.

Smooth Endoplasmic Reticulum

It's main function is to package proteins for transport, synthesise membrane phosolipids, and secrete calcium. It also performs transformation of bile pigments, glycogenolysis (the breakdown of glycogen), and detoxification of different drugs and chemical agents.

Mitochondria

It provides energy to the cell by combining sugar molecules with oxygen to generate carbon dioxide and water.

Amylosplast

It's an organelle present in some plant cells that stores starch.

Druse Crystal

It's a granular type of crystal found in plant vacuoles. It is composed of calcium oxalate and is considered to deter herbivory.

Centrosome

Also known as Microtubule Organising Center, it's a region in the cell where microtubules are produced that perform a variety of functions ranging from transportation to structural support.

Though both plant and animal cell centrosomes play similar roles during cell division, plant cell centrosomes are simpler and do not contain centrioles.

Peroxisomes

These are membrane bound packets of oxidative enzymes that convert fatty acids into sugar and assist chloroplasts in photo-respiration.

Golgi Vesicles

These vesicles transfer proteins and lipids to different parts of the cells.

PLANT CELL MODEL

Following are some simple instructions or plant cell model ideas, to construct a model of a plant cell:

- Enumerate and collect all possible materials required for the model and look out for the different cell parts in a biology textbook.
- Use a heavy sturdy wooden base to construct the base of the model so that it can hold the weight and facilitate presentation of the model.
- Construct the cell nucleus and ensure that it's round in shape. Paint the nucleus in one colour so that it can be identified as a distinct cell part.
- With the help of a dowel rod or thick, straight stick, connect the nucleus to the base of the model. For this, drill a hole into the wooden base, stick the dowel rod into the hole and connect the nucleus to the rod.
- The cell wall and cell membrane of the plant cell model should be rectangular in shape. The outer wall is the cell wall, which should be made of a strong material like hard plastic or wood, and the cell membrane can be made of cellophane or thin plastic. Now attach both the walls to the nucleus and the model base.
- Add various other key components of the plant cell to the model. Like, for chlorophyll, use a piece of green fabric stuffed with cotton. Sew it and attach it to the model.
- Finally, prepare a report on the various plant cell parts and their corresponding functions.

FUNCTIONS OF CELL WALLS IN PLANTS

Plants do not have specialised skeletal systems of the type found in animals; instead, the considerable strength that enables large plants such as the giant *Sequoia* redwood tree to withstand the forces of gravity is provided by the cumulative strength of the walls that surround individual cells. But the load-bearing walls also have to be sufficiently flexible to withstand the large tensile and compressive forces that occur as, for example, the plant sways in the breeze.

The wall is subjected to further forces from within, where osmotic pressure represents another threat to cell and plant integrity. In young cells, the walls must be porous, to allow water, low–molecular weight nutrients and hormones to diffuse freely between cells and tissues, whereas in specialised vascular tissues required for long-distance movement of water and nutrients through the plant, the walls must be impermeable. In most tissues, cell walls are also important for intercellular adhesion.

These various functional requirements are accommodated through the formation of reinforced gel structures consisting of cellulosic microfibrils, which have a high tensile strength, embedded in a gel-like matrix phase that consists predominantly of non-cellulosic polysaccharides. This matrix provides flexibility and some mechanical support for the cell, but its inherent porosity also allows water and other small molecules to diffuse through the wall. As cells of higher plants stop growing and increasingly assume a structural role, their walls thicken, cellulose microfibrils become laminated to increase wall strength and the walls often become encrusted with lignin, which is a large polyphenolic molecule that renders the wall resistant to compressive forces and restricts the passage of small molecules.

At the same time, the wall will 'lock in' the final shape of the cell, and many different cell shapes can be detected in a single section of a plant organ. Smaller amounts of structural and enzymatic proteins and various proteoglycans and glycoproteins are also found in the wall, but these will not be discussed in detail here. In this mature leaf, which has stopped expanding, there are parenchyma cells (p) that have thin walls, fibre cells that have thickened walls, xylem cells that are no longer living and need a wall impervious to the water they transport, and trichome cells (t) that are greatly elongated and require a stiff wall to support them above the surface of the leaf. The cell walls surrounding these cell types play a major role in 'locking in' the final shape of the cell. The midvein (mv) and epidermal layers (e) are also indicated.

In addition to their crucial structural role, cell walls are likely to represent a significant source of metabolisable energy for some plants and may be subject to continual turnover in certain tissues as a result of this function. For example, the walls of the storage cells of seeds are not load bearing but in some cases are extraordinarily thick. Wall polysaccharides in these seeds act as an alternative to starch as a storage macromolecule. Examples include the galactomannans of seeds like fenugreek, carob, ivory nut and guar, as well as pectins in lupin seeds and xyloglucans in nasturtium and tamarind seeds. Grains of many grasses, particularly the economically important cereals such as wheat, maize, rice and barley, contain an endosperm that is rich in starch but also comprises a significant amount of cell wall material that can be used as a source of energy by the germinated grain.

Indeed, it is estimated that up to 18 per cent of metabolisable glucose in the germinated barley grain is released from the (1,3;1,4)-α-D-glucans of the starchy endosperm cell wall. A similar energy reserve function has been proposed for cell wall (1,3;1,4)-α-D-glucans in vegetative tissues of the grasses. Finally, fragments of non-cellulosic polysaccharides of cell walls have been implicated in growth regulation and in signaling pathways associated with a loss of cell wall integrity. Various oligogalacturonan fragments from pectic polysaccharides are bound by receptor-like kinases and can initiate a range of cellular changes, including defence responses that are executed following pathogen attack.

Indeed, one might reasonably ask whether the evolution of the extreme complexity in pectic polysaccharides might have been driven by the potential of these polysaccharides to release an array of different signaling oligosaccharides. If this were true, an obvious next question would be whether other wall polysaccharides, such as the heteroxylans, undertake a similar function, especially in the *Poaceae,* where levels of pectic polysaccharides in the wall are very low.

Alternative Chemistries for Non-cellulosic Polysaccharides

Two major chemical strategies have been used by plants to match non-cellulosic polysaccharide structures to their functional requirements as components of the gel-like matrix phase of walls. Structurally regular polysaccharides with extended conformations of the type commonly found in cellulose and related wall polysaccharides will tend to align and aggregate in aqueous media; the aggregates will be stabilised through the formation of intermolecular hydrogen bonding and hydrophobic interactions between sugar rings.

This will reduce their solubility and enhance their ability to form gels. So, how do plants introduce limited and possibly controllable irregularity into the non-cellulosic wall polysaccharides to limit this tendency to aggregate? In most instances the structurally regular and conformationally extended backbone or main chain that would normally align and aggregate into fibrillar or insoluble complexes is prevented from doing so through the addition of single monosaccharide substituents or short oligosaccharides along the otherwise regular backbone chain. During gel formation in the wall, the presence of some structurally regular 'junction zones' along the polysaccharide chain facilitates localised intermolecular interactions, which, in combination with 'solubilising' non-interacting regions of the chain, allows the polysaccharide to form a gel. An example of this chemical strategy is embodied in the xyloglucans, which consist of a (1,4)-β-glucan main chain. These 'cellulosic' backbone chains do not aggregate because they are substituted with short oligosaccharide chains at the C(O)6 of the main chain (1,4)-β-glucosyl residues.

Similarly, the arabinoxylans of the *Poaceae* consist of a (1,4)-β-xylan main chain that would normally be insoluble and fibrillar in nature but is instead substituted with α-arabinosyl residues. There are also regions of the (1,4)-β-xylan backbone that carry no substituents; these will therefore be capable of intermolecular interactions that will result in limited junction zone formation and will allow gel formation in the wall. The chemical structures of galactomannans and pectic polysaccharides reflect the same strategy.

The second, quite different strategy to achieve wall polysaccharides with extended conformations but with limited tendencies to align and aggregate is exemplified by the relatively recently evolved (1,3;1,4)-β-glucans of the *Poaceae*. Like the xyloglucans, their structure is based upon a conformationally regular 'cellulosic' (1,4)-β-glucan backbone. However, in this alternative strategy, intermolecular aggregation is limited through the introduction of irregularly spaced (1,3)-β-linkages, which result in molecular 'kinks' in the otherwise 'cellulosic' backbone. Provided the (1,3)-β-linkages are spaced at irregular intervals, the asymmetric polysaccharide chains will not aggregate over extended regions.

The (1,3;1,4)-β-glucans of the *Poaceae* have structures based upon a conformationally regular 'cellulosic' (1,4)-β-glucan backbone. Intermolecular aggregation is limited through the introduction of irregularly spaced (1,3)-β-linkages, which result in molecular 'kinks' in the (1,4)-β-glucan backbone. The (1,3)-linked β-glucosyl residues are shown in red, whereas the (1,4)-β-glucosyl residues are shown in blue. Relatively high or relatively low ratios of cellotriosyl/cellotetraosyl units indicate a predominance of cellotriosyl or cellotetraosyl residues, respectively.

If one or another of the oligosaccharide units predominates, the overall polysaccharide chain becomes more regular in shape and hence less soluble because of chain alignment. Random arrangement of a mixture of the trisaccharide and the tetrasaccharide units means that the chains will not aggregate over extended regions and can remain in aqueous solution, as depicted in the lower diagram. Such differences in physicochemical properties could be used to the plant's advantage in meeting different functional requirements in different cells or at different stages of development.

Molecular Mechanisms for Wall Polysaccharide Biosynthesis

The generally accepted mechanism of wall polysaccharide synthesis involves enzymes that add glycosyl residues from activated sugar nucleotide donors to the elongating polymeric chain. Two major classes of these glycosyltransferases have been recognised. One group, which is often referred to as the polysaccharide synthase group, is believed to catalyse the iterative addition of glycosyl residues to the end of the elongating polysaccharide chain, without releasing the chain between successive glycosylation events. It has been generally assumed that the incoming glycosyl residue is transferred from the sugar nucleotide donor substrate to the non-reducing end of the polysaccharide, although it has also been suggested that plant cell wall polysaccharide biosynthesis might involve addition of glycosyl residues to the reducing end of the elongating chain.

The iterative synthase enzymes are assumed to catalyse the polymerisation not only of cellulose but also of the main chain of non-cellulosic wall polysaccharides; these enzymes include the (1,4)-βmannan synthase that polymerises the (1,4)-β-mannan backbone of mannans, galactomannans and galactoglucomannans and the (1,4)-β-glucan synthase that synthesises the backbone of xyloglucans. These enzymes are encoded by members of the cellulose synthase (*CesA*) and cellulose synthase-like (*Csl*) gene superfamily and are members of the GT2 group of glycosyltransferases. The GAUT enzymes are also believed to be iterative glycosyltransferases that synthesise the (1,4)-β-galacturonan backbone of pectic polysaccharides, but the GAUT galacturonyl transferases belong to the GT8 group of enzymes, and their structures are likely to be quite distinct from the GT2 group involved in the synthesis of β-linked polysaccharide chains.

The second group of glycosyltransferases is believed to add single glycosyl substituents to the main chain residues or to add short oligosaccharide chains that are appended to the main chain through single catalytic encounters with the nascent polysaccharide chain. Examples of this group of glycosyltransferases include the arabinosyl transferases that add á-arabinofuranosyl or β-glucuronyl residues to the (1,4)-β-xylan backbone of arabinoxylans, the xylosyl transferases that add (1,6)-β-xylosyl residues to the (1,4)-β-glucan backbone of nascent xyloglucan molecules and the galactosyl transferases that add (1,6)-β-galactopyranosyl residues to the backbone of (1,4)-β-mannans or (1,4)-β-glucomannans.

Despite the best efforts of biochemists over many decades, there are still a number of fundamental questions about the biosynthetic process that we have not answered. The mechanism of chain initiation is not known, and potential primer molecules for wall polysaccharide biosynthesis have not been convincingly identified. Similarly, the mechanisms controlling chain termination and thus the overall length of the polysaccharide chain have not been elucidated. Do polysaccharide chains simply 'fall off' the synthase enzymes when they get to a certain length or when the enzyme itself begins to be turned over? Do the hydrolytic enzymes that have been increasingly recognised in tissues in which net synthesis of wall polysaccharides is occurring hydrolyse the elongating chain and release the nascent polysaccharide from the synthase enzyme? Might decreasing concentrations of glycosyl donor substrates lead to termination of the chain? It was recently suggested that a terminal oligosaccharide detected at the reducing end of certain heteroxylans might be involved in chain termination.

Whichever of these or other possibilities might apply, plant cell wall polysaccharide synthesis does not appear to be guided by a template, and it is difficult to envision how chain termination could be anything but an imprecise process. It follows that the resultant polysaccharide chain length will be heterogeneous, as is usually observed. This is not to say, however, that plant cells do not have some control over wall polysaccharide chain length, because clear differences are apparent, and in some cases chain lengths fall within reasonably narrow limits.

A second potential level of imprecision is afforded by the glycosyltransferases that add the various glycosyl units to the backbone of the wall polysaccharide. We do not know whether the addition of the substituent is a precise process or not, insofar as there are examples of relatively regular spacing of substituents along the chain and other examples in which substituents are, in statistical terms, asymmetrically distributed. In the former case, one might expect that the backbone synthase and the substituting glycosyltransferase would have to operate as a highly coordinated enzyme complex.

On the other hand, if the main chain synthase and glycosyltransferase were operating independently, it is again difficult to see how main chain substitution could be a precise process with respect to the extent and spacing of added substituents. Equally plausible is the possibility that an enzyme complex could mediate the generation of a polysaccharide in which every main chain residue was substituted or in which every second or third main chain residue was substituted. This would result in a relatively homogeneous polysaccharide, but postsynthetic removal of substituents at regular intervals, as has been described for arabinoxylans, or asymmetrical postsynthetic modifications such as methyl esterification, as has been described for some pectic polysaccharides, could introduce heterogeneity into a polysaccharide that was initially homogeneous with respect to its fine structure.

Against this background of heterogeneity that will arise generally as a result of biosynthetic mechanisms, the specific types of heterogeneity actually observed in the major groups of non-cellulosic wall polysaccharides are summarised in the sections below.

Heterogeneity in Pectic Polysaccharides

Pectins are a major component of the primary cell wall, particularly of dicotyledonous plants. Pectins contribute to the mechanical strength, porosity, adhesion and stiffness of the cell wall and have been implicated in many processes, including wall slippage, extension and intercellular signaling. In addition, pectic polysaccharides are a source of signaling molecules.

Pectins are extremely complex molecules that include a number of structural classes, such as homogalacturonan, xylogalacturonan, apiogalacturonan and rhamnogalacturonan. The galacturonan chains can chelate calcium ions, apiosyl residues can coordinate borate ions, and there is good evidence that covalent linkages form between different structural classes of the pectic polysaccharide group. Methylated and acetylated residues are often found on pectin and indeed on several other polysaccharides and can profoundly influence their structures and physicochemical properties.

Considering this complexity, it is not unexpected that the level of heterogeneity in the pectic polysaccharides is high. Here we will explain just one example of how structural heterogeneity is generated and how this alters the physicochemical and functional properties of the cell wall matrix. A common mechanism facilitating local changes in the properties of pectic polysaccharides is the degree to which their polygalacturonate residues are methyl esterified. This involves heterogeneity both within a chain and between chains and can now be characterised analytically.

Methyl esterification will neutralise the charge on galacturonyl residues and thereby abolish the ability of the negatively charged galacturonyl residues to interact with calcium ions. Specific monoclonal antibodies, which can differentially detect pectins with particular methylation profiles, show spatial heterogeneity around a single higher-plant cell with respect to the pattern of pectin esterification, which may be linked to cell adhesion. There are a number of pectin methyl esterases that selectively remove methyl groups from homogalacturonan in either a blockwise or a non-blockwise fashion.

Where methyl groups are removed in a blockwise manner, an exposed region of the polysaccharide that is able to form calcium-mediated cross-links is created; these cross-links contribute to the 'stickiness' required for cell adhesion. Where the removal of ester groups is non-blockwise, the homogalacturonan appears to be less 'sticky', and such junction zones cannot form so readily. As one might expect, this particular form of homogalacturonan is found in regions where cells separate.

Heterogeneity in Heteroxylans

Heteroxylans have a (1,4)-β-xylan backbone that can be substituted with single α-arabinofuranosyl, α-glucuronosyl or 4-methyl-α-glucuronosyl residues. These residues can be linked to the $C(O)_3$ or $C(O)_2$ atoms of the backbone xylopyranosyl residues, or in some cases the xylopyranosyl residues can be substituted at both $C(O)_2$ and $C(O)_3$. Short oligosaccharide substituents are also found on some heteroxylans and, in the grasses in particular, á-arabinofuranosyl substituents can be esterified with hydroxycinnamic acids such as ferulic acid and *p*-coumaric acid. Two carbohydrate-binding modules and two monoclonal antibodies have been used to locate specific types of xylans both within intact cell walls and in solubilised wall material. Both classes of probe revealed distinct types of heteroxylans, which differ in both structure and charge and are present in both primary and secondary walls.

Similarly, in the endosperm of the grasses, wall arabinoxylans digested with (1,4)-β-xylan endohydrolase release a population of oligomeric arabinoxylosides that provide an enzymatic fingerprint of the polysaccharide's chemical structure. Examination of the arabinoxylan fingerprint from wheat endosperm walls showed variation in the structure and amount of this polysaccharide both spatially and temporally through grain development. The results were confirmed through Fourier transform infrared microspectroscopy (FT-IR), which showed that arabinoxylans in developing grain contained more backbone xylosyl residues disubstituted with arabinosyl residues than did arabinoxylans in mature grain and that the structure of the arabinoxylan was dependent upon its position in either prismatic or central cells.

Heterogeneity in Xyloglucans

Xyloglucans are thought to be particularly important in the primary walls of many plants because of their ability to non-covalently cross-link cellulose microfibrils. Their structure is based on a cellulosic (1,4)-β-glucan backbone substituted at $C(O)_6$ with α-linked xylosyl residues, typically at about 70 per cent of the backbone glucosyl residues. However, further substitution of xylosyl residues is required to prevent xyloglucan self-association, which interferes with binding to cellulose. A typical basic substructure involves xylosyl substituents on three consecutive glucosyl backbone residues, followed by an unsubstituted glucosyl residue. The availability of endoglucanase enzymes that depolymerise the xyloglucan chain at unsubstituted glucosyl residues allows the isolation of oligosaccharides for detailed analysis.

A wide range of carbohydrate and selective acetyl substituents on the basic xyloglucan structure have been characterised in this way. Most studies of partial glucanase digests are consistent with an essentially random arrangement of oligosaccharide substructures, but this may not always be the case. One promising new approach to characterising the heterogeneity of xyloglucans is the use of mass spectrometry to define glucanase-derived oligosaccharides on very small sample sizes. The selectivity of specific endoglucanase digestion, coupled with the sensitivity of mass spectrometric detection, allows not only the profiling of small amounts of isolated xyloglucan polymers but also the profiling of xyloglucans present within specific cell types or in isolated organelle fractions. Furthermore, *in situ* analysis of xyloglucan oligosaccharide substructures in walls can be carried out at high resolution by aiming the ionisation laser of a MALDI-TOF mass spectrometer at desired locations within enzyme-treated tissues.

Although mass spectral methods depend to some extent on prior knowledge of xyloglucan structure and cannot be easily applied to polysaccharides such as arabinoxylans, which consist only of pentose sugars, the method has potential to open up the dissection of location-specific heterogeneity in xyloglucan structures, especially if the method can be complemented with antibodies directed against an array of xyloglucan substructures.

Heterogeneity in Mannans

The cell wall polymer family for which fine structure has been most precisely defined and reconciled with biosynthesis processes is the galactomannans, members of which have a (1,4)-β-mannan backbone along which some but not all 6-C(OH) positions are substituted with α-galactose. Mannose/galactose ratios vary from about 1:1 to as high as 20:1. On the basis of enzyme digestion and quantitative analysis of digestion-resistant oligosaccharides, it has been shown that second-order Markov chain statistics are sufficient to describe the fine structure of a range of galactomannans.

For a processive biosynthesis mechanism in which galactosyl transferase activity is coupled with backbone mannan synthase, a second-order Markov chain is consistent with the galactomannan subsite recognition sequence of the galactosyltransferase encompassing at least three backbone mannosyl residues—that is, the site of substitution and the two preceding residues towards the reducing end of the growing galactomannan chain. Datasets from three different plant species generated three distinctly different sets of probabilities and thus different galactosyl-substitution patterns.

Subsequently, heterologously expressed galactosyl transferases in transgenic tobacco resulted in the production of galactomannan with Markov probability values characteristic of the original source of the galactosyl transferase. The Markov statistics describing polymer structure also allow a precise prediction of the probability of unsubstituted mannan segments of defined length, and these have been shown to be of consequence in the formation of interpolymeric junction zones that can form the basis for three-dimensional gel-like matrices of the type thought to be present within plant cell walls.

Heterogeneity in (1,3;1,4)-β-glucans

The (1,3;1,4)-β-glucans, which are abundant in walls of the *Poaceae*, represent one example in which heterogeneity in fine structure appears to be essential for function in the wall and is therefore built into the (1,3;1,4)-β-glucan during biosynthesis. Detailed structural analyses of the (1,3;1,4)-β-glucans demonstrate that randomness can be recognised at one level, whereas other structural features of the molecule are non-random. How can this be?

The water-soluble (1,3;1,4)-β-glucan from barley endosperm consists of about 72 per cent (1,4)-β-glucosyl residues and 28 per cent (1,3)-β-glucosyl residues. The (1,3)-β-glucosyl residues are always flanked on both their reducing end and non-reducing end sides by (1,4)-β-glucosyl residues; two or more adjacent (1,3)-β-glucosyl residues are not found. In this respect the fine structure is non-random, because if the (1,3)-β-glucosyl residues and (1,4)-β-glucosyl residues were randomly distributed, on statistical grounds one would expect to find significant numbers of two or more adjacent (1,3)-β-glucosyl residues.

The (1,4)-β-linkages of most (1,3;1,4)-β-glucans are arranged predominantly in groups of two or three adjacent linkages. In the water-soluble (1,3;1,4)-β-glucan from barley grain, this arrangement accounts for about 90 per cent of the polysaccharide. The remaining 10 per cent consists of longer blocks of from 4 up to 12–15 adjacent (1,4)-β-glucosyl residues. If we focus on the groups of two and three adjacent (1,4)-β-glucosyl linkages, (1,3;1,4)-β-glucans can be viewed as (1,3)-β-linked copolymers of cellotriosyl and cellotetraosyl residues in which the ratio of cellotriosyl (DP3) to cellotetraosyl (DP4) units varies from 1.5 to 4.5, depending on the source of the (1,3;1,4)-β-glucan. Markov chain analyses were used to investigate the arrangement of the cellotriosyl (DP3) to cellotetraosyl (DP4) units and showed that the cellotriosyl (DP3) to cellotetraosyl (DP4) units were arranged at random along the (1,3;1,4)-β-glucan chain.

The structure of a water-soluble barley (1,3;1,4)-β-glucan was investigated by treating it as a Markov chain. The polysaccharide consists mostly of cellotriosyl and cellotetraosyl units linked by (1,3)-β-linkages. This analysis was used to explore whether the addition of a particular unit oligosaccharide structure to the elongating chain depended on the adjacent unit oligosaccharide in the chain. Through controlled enzymatic hydrolysis, such that most of the products represented two units of the polysaccharide chain, it was possible to quantitate the amount of hexasaccharide (which consisted of two cellotriosyl units), heptasaccharide (which consisted of one cellotriosyl unit and one cellotetraosyl unit) and octasaccharide (which consisted of two cellotetraosyl units).

Applying this quantitative measure of near-neighbour analysis in the Markov chain would give a value for ρ of about 1.0 for blocks of adjacent cellotriosyl units and adjacent cellotetraosyl units, with values near –1.0 if the polysaccharide consisted of alternating cellotriosyl and cellotetraosyl units and values close to zero if the cellotriosyl and cellotetraosyl units were randomly arranged. The experimentally determined values for ρ for two different but related (1,3;1,4)-β-glucan preparations were –0.003 and 0.050.

The data therefore indicated that the (1,3)-β-linked cellotriosyl and cellotetraosyl units were arranged essentially at random along the polysaccharide chain. *p*, the probability that a cellotriosyl unit is followed by another cellotriosyl unit. *q*, the probability that a cellotetraosyl unit is followed by a cellotriosyl unit. The heterogeneity in fine structure imparted by this chemistry carries important consequences for the physicochemical and hence functional properties of the (1,3;1,4)-β-glucan.

Thus, mid-level ratios of cellotriosyl/cellotetraosyl units, say 1.5:2.5, would meet the functional requirements of a porous, gel-like wall matrix, whereas much higher or lower ratios would characterise a more conformationally regular, less soluble (1,3;1,4)-β-glucan that would have increased capacity to aggregate with other (1,3;1,4)-β-glucan molecules and possibly with cellulose and other wall polysaccharides such as heteroxylans. The wide adoption of (1,3;1,4)-â-glucans in the *Poaceae* might be attributable to the ability of these species to synthesise (1,3;1,4)-β-glucans with mid-range ratios of cellotriosyl/cellotetraosyl units.

Biological models for the synthesis of the (1,3;1,4)-β-glucans must account for their fine structural features, three of which challenge traditional cellular and biochemical views of how wall polysaccharides are synthesised.

These are:

- The occurrence of single, isolated (1,3)-β-glucosyl residues amongst the predominating (1,4)-β-glucosyl residues,
- The random arrangement of cellotriosyl units and cellotetraosyl units along the chain
- The incorporation of longer blocks of up to 12–5 adjacent (1,4)-β-glucosyl residues in the molecule.

A two-phase assembly system has been proposed to account for the structural features of (1,3;1,4)-β-glucans. As illustrated diagrammatically in Figure, the first phase involves the assembly of a population of cellodextrins, which are linked to a Golgi membrane associated macromolecule. It is envisioned that a population of (1,4)-β-oligoglucosides is synthesised by the cellulose synthase–like CslF or CslH enzymes, some of which are located in the Golgi. After transfer of the membrane-bound (1,4)-β-oligoglucosides to the plasma membrane, the second phase of the model suggests that the (1,4)-β-oligoglucosides are taken at random by another enzyme, possibly a callose synthase, a xyloglucan endotransglycosylase or even another CslF isoenzyme, and linked through (1,3)-β-linkages.

The model is consistent with the cellular location of the final (1,3;1,4)-β-glucan polysaccharide. During extensive immunochemical examination of (1,3;1,4)-β-glucan synthesis in a range of different cell types, the polysaccharide could not be detected in the Golgi, even where its concentration in the wall was very high. The Golgi-located cellodextrins proposed to represent the first phase of (1,3;1,4)-β-glucan synthesis would not be detected by the (1,3;1,4)-β-glucan antibodies used. However, another group has recently disagreed and has concluded that some polymeric (1,3;1,4)-β-glucan is located in Golgi-derived vesicles of maize coleoptiles. We predict that these cell biological aspects of cell wall polysaccharide synthesis will remain the topic of debate for some time.

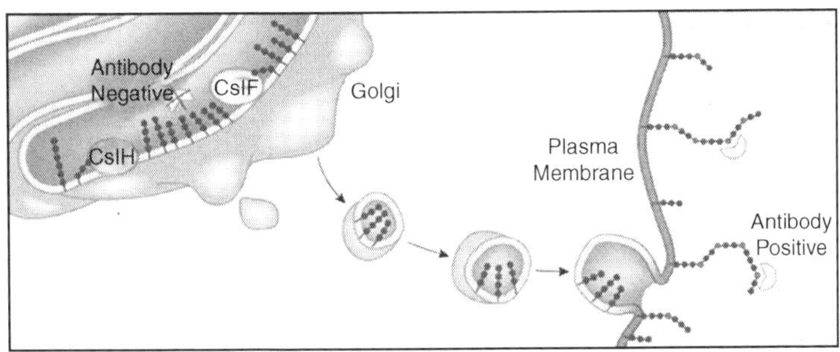

Fig. An Alternative, Two-phase Mechanism for Non-cellulosic Polysaccharide Assembly.

Non-cellulosic polysaccharides of the matrix phase in plant cell walls are generally considered to be synthesised in their entirety in the Golgi apparatus, from which they are transported via Golgi-derived vesicles to the plasma membrane. Fusion of the vesicles with the plasma membrane results in the deposition of the polysaccharides into the periplasmic space and their eventual incorporation into the wall. In an alternative theory, it has been proposed that oligosaccharides might be assembled in the Golgi, where they remain attached to a recyclable membrane lipid (or protein). This would be consistent with early but not generally accepted indications that membrane-associated sitosterol or other sterols might act as intermediates in cellulose synthesis.

The oligosaccharides would subsequently be polymerised by a plasma membrane–bound transferase that would deposit the nascent polysaccharide chain directly into the wall. Callose synthase and xyloglucan endotransglycosylases were suggested as candidate oligosaccharide polymerases. In this diagram, a model for the potential assembly of (1,3;1,4)-β-glucans is presented, through which a population of different-length cellodextrins are synthesised in the Golgi by CslF and CslH enzymes.

These cellodextrins, which the (1,3;1,4)-β-glucan antibody does not recognise, would then be transported to the plasma membrane via Golgi-derived vesicles. Upon fusion with the plasma membrane, the cellodextrins are released to allow the second stage os assembly to occur where they are joined by (1,3)-β-linkages.

This yields the final polysaccharide now recognised by the (1,3;1,4)-β-glucan antibodies. It has been postulated that an assembly system similar to this might also apply to other non-cellulosic polysaccharides of plant cell wall. Although the theory of two-phase assembly of wall polysaccharides in the Golgi and at the plasma membrane was developed initially to account for experimental data related not only to the cell biology of (1,3;1,4)-β-glucan biosynthesis but also to the inbuilt heterogeneity in their fine structure, the model might be equally applicable to the synthesis of other wall polysaccharides. Xyloglucans have a strong repetitive element in their structures and might be considered a polymer of xylosylated tetraglucoside units. Pectic polysaccharides contain some repeating elements, and extracellular arabinogalactan proteins are believed to be made up of repeating oligosaccharide units.

A Potential Role for Hydrolases in Creating Heterogeneity

An emerging theme in cell wall biology is the presence of polysaccharide hydrolases in tissues in which walls are being deposited and there is net synthesis of constituent wall polysaccharides. One example is the presence of (1,4)-β-glucan endohydrolases (cellulases) such as Korrigan in tissues in which cellulose synthesis is occurring. The role of the cellulase has not been defined unequivocally, but it may be involved in the release of nascent cellulosic chains from the cellulose synthase enzyme complex, or it might be involved in the trimming of non-crystalline cellulose from the surface of microfibrils.

The latter possibility is consistent with observations that cellulose crystallinity is decreased in *Arabidopsis* lines that carry a silencing mutation in the *kor* gene. Molecular heterogeneity in heteroxylans might be generated by partial hydrolysis of the nascent polysaccharide rather than during the biosynthetic process itself. Newly synthesised arabinoxylans in maize and barley coleoptiles are usually highly substituted. Indeed, most if not all the xylosyl residues of the (1,4)-β-xylan backbone are initially substituted with arabinosyl residues. However, as the wall matures the degree of substitution decreases, and this has been attributed to the activity of arabinoxylan arabinofuranohydrolases, which can remove arabinosyl residues from the (1,4)-β-xylan backbone.

Irrefutable emerging evidence points to high levels of chemical, and hence structural, heterogeneity in the polysaccharide constituents of plant cell walls. The heterogeneity can be measured between walls in different parts of the same plant, between walls in a single tissue section and between regions within a single cell wall. In addressing the 'big questions' in plant cell wall biology in the immediate future, we will need to ensure that our explanations adequately rationalise heterogeneity as a crucial component of wall structure and function. Of the major unanswered questions facing cell wall biologists, the unequivocal identification of enzymes involved in the biosynthesis of the backbones of xyloglucans and heteroxylans is particularly pressing. Given the evidence for the participation of multienzyme complexes in cellulose biosynthesis, it will be important to investigate the highly likely possibility that the synthesis of non-cellulosic wall polysaccharides also requires the coordinated action of several enzymes and other proteins. The latter question becomes even more intriguing in the case of substituted wall polysaccharides, in which enzymes involved in the synthesis of the backbone might form a complex with enzymes that simultaneously add monosaccharide or oligosaccharide substituents to the backbone. Might hydrolases also form a part of such complexes?

The recent observation that different isoforms of a single enzyme family can direct the synthesis of wall polysaccharides with different structures and thus different physicochemical properties raises the question of what the molecular mechanisms used by enzymes to synthesise wall polysaccharides in plants are. To thoroughly and accurately describe these mechanisms, three-dimensional structures of the synthases will be required, but at this stage there are no crystal structures of plant polysaccharide synthases of the GT2 group in the public databases.

The conventional view of the cellular steps involved in non-cellulosic wall polysaccharide assembly has also been challenged, although proposed alternative pathways are supported largely by indirect data. It is likely that new technologies will enable us to investigate alternative models of wall polysaccharide assembly at the cellular level in the immediate future; the results of these investigations will be potentially transformational in terms of the way we view the cell biology of wall metabolism in plants. Similarly, emerging *in situ* methods will enable us to define in more detail the nature of polysaccharide-polysaccharide, polysaccharide-protein and polysaccharide-lignin interactions in the wall, to define whether these interactions involve covalent or non-covalent linkages and to identify the enzyme systems responsible for mediating the interactions. It has been known for some years that oligosaccharides released from wall polysaccharides by the action of microbial enzymes can act as signal molecules and, in particular, signal a loss of cell wall integrity that elicits a range of plant defence responses to pathogen attack. Given the role of the cell walls of both the microorganism and the plant as contact points during their first encounter, the question of cell wall integrity and oligosaccharide signaling will continue to be of importance in describing how plants resist biotic stress. The structural complexity and heterogeneity of pectins are consistent with a role for these polysaccharides in these processes, but are there other polysaccharides in the walls of the *Poaceae*, where pectin levels in walls are often very low, that assume a similar role?

CELL MODEL OF PLANT

Biotechnology is name given to the methods and techniques that involve the use of living organisms like bacteria, yeast, plant cells etc or their parts or products as tools (for example, genes and enzymes). They are used in a number of fields: food processing, agriculture, pharmaceutics, and medicine, among others. Plant tissue culture can be defined as culture of plant seeds, organs, explants, tissues, cells, or protoplasts on nutrient media under sterile conditions. The science of plant tissue culture takes its roots from path breaking research in botany like discovery of cell followed by propounding of cell theory. In 1839, Schleiden and Schwann proposed that cell is the basic unit of organisms. They visualised that cell is capable of autonomy and therefore it should be possible for each cell if given an environment to regenerate into whole plant. Based on this premise, in 1902, a German physiologist, Gottlieb Haberlandt developed the concept of *in vitro* cell culture.

He isolated single fully differentiated individual plant cells from different plant species like palisade cells from leaves of Laminum purpureum, glandular hair of Pulmonaria and pith cells from petioles of Eicchornia crassiples etc and was first to culture them in Knop's salt solution enriched with glucose. In his cultures, cells increased in size, accumulated starch but failed to divide. Therefore, Haberlandt's prediction failed that the cultured plant cells could grow, divide and develop into embryo and then to whole plant. This potential of a cell is known as totipotency, a term coined by Steward in 1968.

Despite lack of success, Haberlandt made several predictions about the requirements in media in experimental conditions which could possibly induce cell division, proliferation and embryo induction. G Haberlandt is thus regarded as father of tissue culture. Taking cue from Haberlandt's failure, Hannig chose embryogenic tissue to culture. He excised nearly mature embryos from seeds of several species of crucifers and successfully grew them to maturity on mineral salts and sugar solution. In 1908, Simon regenerated callus, buds and roots from Poplar stem segments and established the basis for callus culture. For about next 30 years, there was very little further progress in cell culture research. Within this period, an innovative approach to tissue culture using meristematic cells like root and stem tips was reported by Kolte and Robbins working independently.

All these research attempts involving culture of isolated cells, root tips or stem tips ended in development of calluses. There were two objectives to be achieved before putting Haberlandt's prediction to fruition. First, to make the callus obtained from the explants to proliferate endlessly and second to induce these regenerated calluses to undergo organogenesis and form whole plants. It was in 1930s, when progress in plant tissue culture accelerated rapidly owing to an important discovery that vitamin B and natural auxins were necessary for the growth of isolated tissues containing meristems.

This breakthrough came from White who reported that not only could cultured tomato root tips grow but could be repeatedly subcultured to fresh medium of inorganic salts supplemented with yeast extract. He later replaced YE by vitamin B namely pyridoxine, thiamine and proved their growth promoting effect.

DESCRIPTION

Plant cells are the cells found in plants comprising sub cellular organelles. Following is a plant cell model for kids that will give your child a visual of the plant cell and a better understanding of its various parts.

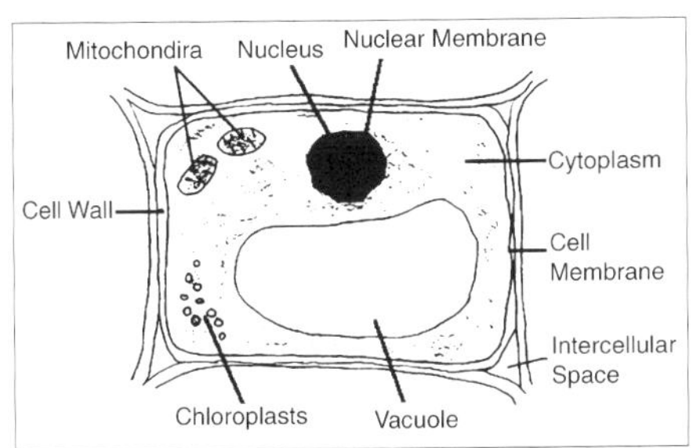

The cells are the basic units of life that work together to perform life sustaining functions in both animal and plant worlds. Plant cells are eukaryotic cells having thick and rigid cell walls.

Eukaryotic cells are cells that contain complex structures enclosed within membranes called nucleus or nucleus envelop, within which the genetic material is present. These cells are present in almost all living organisms including animals, plants and fungi.

STRUCTURE AND FUNCTION OF PLANTS

The digestibility of plants (*i.e.*, how easy they are to digest) affects the efficiency of animals' digestive processes, and is related to the structure of plant cells and tissues.

DIGESTIBILITY OF PLANTS

Animals can easily digest the contents of plant cells, but not their cell walls. All plant cells have an outer cell wall composed of cellulose or lignin.. This reduces the amount of energy animals can extract from plants, because most animals lack the specific enzymes that would allow them to digest cellulose. Fibre is the amount of cellulose and lignin present in the plant. In general, older and woodier plants have higher levels of cellulose and lignin. Lignin cannot be digested no matter how long it remains in the digestive system.

In addition, lignin interferes with gut microbes that do have the necessary enzymes to digest cellulose, because it both acts as a physical barrier to digestion, and contains chemical bonds that cannot be broken down by normal stomach microbial flora. 'Digestibility" describes how readily the plant can be digested and its energy released for use by an animal.

The digestible part of a plant includes the cell contents and the small amount of fibre that can be broken down. The amount of energy available to an animal eating a plant is determined by the proportion of fibre to cell contents in that plant.

Digestibility of Plant Structures

Grasses normally have higher fibre content than legumes such as clover. The amount of fibre in leaves of grass is twice that of legume leaves, and grass leaves are harder to digest than those of legumes. The leaves of grasses typically have a midrib made of lignin to provide support for the leaf, which adds to the higher fibre levels and lower digestibility of the leaves. The stems of most plant species have greater fibre levels compared to the leaves, and grass stems usually contain more fibre than legumes. Digestibility not only declines down the stem, but also declines more rapidly than leaf digestibility with increasing plant age (Buxton and Redfearn, 1997).

Digestive Efficiency of Ruminants

Ruminants are better than non-ruminants at digesting high-fibre diets. Leaves have a shorter rumen retention time than stems due to both faster rates of fibre digestion and higher rates of passage of undigested material. Ruminants can digest 40-50 per cent of legume fibre and 60-70 per cent of grass fibre (Buxton and Redfearn, 1997). Small particles of food are digested faster than large particles because they have more surface area exposed relative to the total food volume. This is why ruminants regurgitate and re-chew (ruminate) their food after eating, to reduce the size of the food particles. Because high fibre plants are harder to break down, ruminants spend more time regurgitating and chewing grasses than legumes and more time chewing mature than immature plants.

Digestive Efficiency of Hindgut Fermenters

Hindgut fermenters such as horses do not re-chew their food. They rely on eating more food to extract enough nutrients. This means that at certain times of the year, non-ruminants must eat lower quality feed in order to sustain their energy requirements. So when food becomes limited, animals such as horses and pandas must eat whatever is available, while cattle may pick and choose. In both ruminants and non-ruminants efficiency of digestion drops off as dietary fibre reaches certain levels. Above 60 per cent fibre, the efficiency of digestion in both cattle and horses drops to the point whereby not enough energy can be extracted from the food to make it worth eating. This is about the same level of fibre as the wood of many trees.

Environmental Effects of Low Digestibility

Due to the comparatively low digestibility of grass and other vegetative matter, ruminants and hindgut fermenters must consume proportionally more food than carnivores or omnivores. The high food intake and low nutrient extraction of ruminants and hindgut fermenters means that they concentrate and excrete large quantities of plant nutrients such asphosphorus and nitrogen. This leads to problems such as pollution and eutrophication of lakes and rivers when large numbers of cattle are concentrated on a comparatively small amount of farmland. Phosphorus and nitrogen from cattle effluent is washed in to lakes and waterways and causes blooms of algae that reduces water quality.

CELLULOSE AND LIGNIN

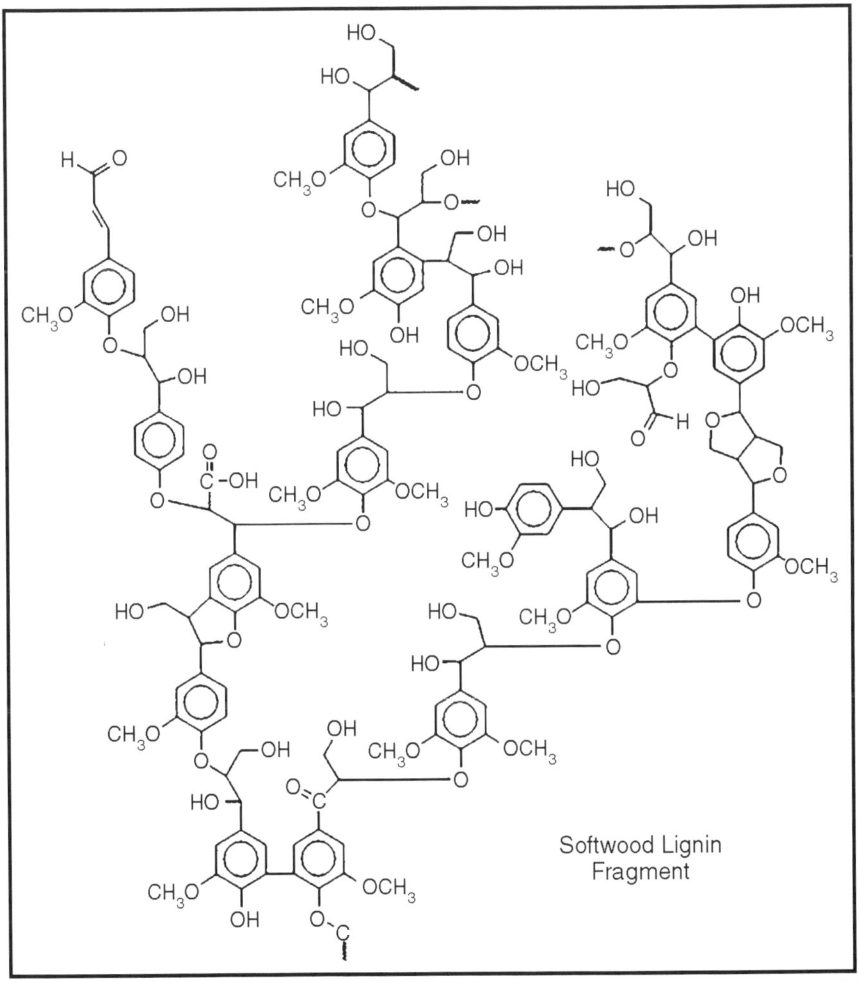

Softwood Lignin Fragment

The plant cell wall surrounds the cell membrane. It is made up of multiple layers of cellulose which are arranged into primary and secondary walls. Cellulose is the most common organic compound on Earth. About 33 per cent of all plant matter is cellulose - the cellulose content of cotton is 90 per cent and wood is 50 per cent cellulose. The cell walls of all vascular plants also contain a polymer called lignin. Lignin is water-resistant. It reinforces cell walls, keeping them from collapsing. This is particularly important in the xylem, because the column of water in the hollow xylem cells is under tension (negative pressure) and without the lignin reinforcement the cells would collapse.

Cellulose

Cellulose is a polymer made of repeating glucose molecules attached end to end (Thus cellulose is an example of a polysaccharide.). A cellulose molecule may be from several hundred to over 10,000 glucose units long. Cellulose from wood pulp has typical chain lengths between 300 and 1700 units; cotton and other plant fibres have chain lengths ranging from 800 to 10,000 units (Klemm *et al.* 2005).

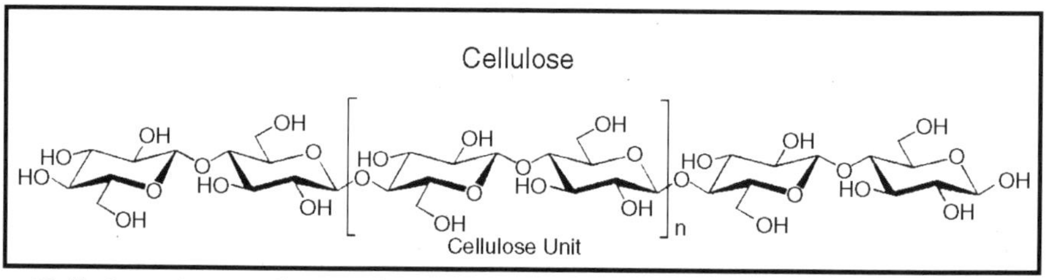

Cellulose

Cellulose Unit

Cellulose is similar in form to complex carbohydrates like starch and glycogen. These polysaccharides are also made from multiple subunits of glucose. The difference between cellulose and other complex carbohydrate molecules is how the glucose molecules are linked together. In addition, cellulose is a straight chain polymer, and each cellulose molecule is long and rod-like. This differs from starch, which is a coiled molecule. A result of these differences in structure is that, compared to starch and other carbohydrates, cellulose can not be broken down into its glucose subunits by any enzymes produced by animals.

Lignin

Lignin provides the mechanical support for stems and leaves and supplies the strength and rigidity of plant walls. Lignin provides the structural strength needed by large trees to reach heights in excess of 100 m. Without lignin these trees would collapse on themselves. Also, lignin along with other cell wall constituents provides resistance to diseases, insects, cold temperatures, and other stresses. Lignin plays a crucial part in conducting water in plant stems. The structure of lignin has not been properly determined as it usually fragments upon extraction and there appears to be no consistent structure to it. The polysaccharide components of plant cell walls are highly hydrophilic and thus permeable to water, whereas lignin is more hydrophobic. The crosslinking of polysaccharides by lignin is an obstacle for water absorption to the cell wall. Thus, lignin makes it possible for the plant's vascular tissue to conduct water efficiently. Lignin is present in all vascular plants, but not in bryophytes, supporting the idea that the original function of lignin was restricted to water transport.

PLANT CELLS AND TISSUES

Cells are the smallest functional units of life. A living organism may comprise a single cell, *e.g.*, an alga; or it may be a multicellular organism made up of billions of cells, *e.g.*, a kauri tree. An individual plant contains many different cell types, each adapted to perform a particular function. However, each living plant cell is made up of the same basic components: a cell wall, plasma membrane, nucleus, and mitochondria and other organelles.

Plant Cell Structure

Each plant cell is self-contained and at least partially self-sufficient. This is because plant cells have a number of organelles that perform functions such as storing food, processing waste, fixing and releasing chemical energy, and copying their genetic material. The nucleus, the mitochondrion, the plastids, the Golgi apparatus, and the endoplasmic reticulum are all examples of organelles.

Some organelles, such as mitochondria and chloroplasts, have their own genetic material separate from that found in the nucleus of the cell. Such organelles are thought to have originated through the process of endosymbiosis. Other structures such as the cell wall and the cell membrane serve to protect the cell and maintain the shape, structure, and functioning of the cell.

The Plant Cell Wall

Unlike animal cells plant cells have an outer cell wall composed of polysaccharides, in particular cellulose. Cell walls provide rigidity and mechanical support, maintain cell shape and the direction of cell growth. The cell wall also prevents expansion when water enters the cell.

Cell Wall Structure

The cell wall of a plant cell is made up of primary and secondary cell walls. When one cell divides into two, a primary cell wall forms around each two new cells. The primary cell wall is laid down just outside the cell membrane, and is flexible, allowing the new cells to grow. This new cell wall is mostly made of cellulose molecules, arranged into thin hair-like strands called microfibrils. The microfibrils are arranged in a meshwork pattern along with other components such as hemicellulose, glycans and pectins, which link them together and help strengthen the cell wall. The secondary cell wall is constructed between the plasma membrane and the primary cell wall after the cell has finished growing. It is made by laying down successive layers of cellulose microfibrils and lignins. Mature xylem cells are heavily lignified - they make up the 'wood' of woody plants (Raven *et al.* 2005).

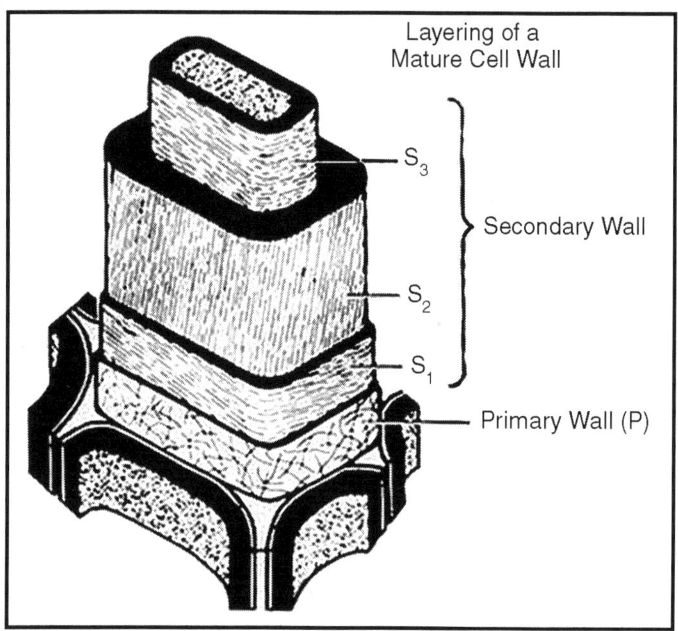

Cell Plasma Membrane

Every living cell is bounded by an outer membrane called the cell plasma membrane, or plasmalemma. The cell membrane separates the cell's contents from the surrounding environment and controls the movement of materials into and out of the cell.

The cell membrane is made up of two phospholipid layers (a bi-layer) which spontaneously arrange so that the hydrophobic "tail" regions are protected from the surrounding water, causing the morehydrophilic "head" regions to be pointed towards either the cytoplasm or the exterior of the cell. Substances such as amino acids, nucleic acids, carbohydrates, proteins, and many ions are unable to diffuse across the membrane but must enter or leave the cell through active transport, under the control of the cell membrane.

This involves some of the many different proteins and lipids embedded in the plasma membrane. Other proteins are involved in other cellular processes such as cell linkage, ion channel flow and cell communication. The plasma membrane also serves as the attachment point for both the intracellular cytoskeleton (found in all eukaryote cells) and the cell wall.

Cytoplasm

The cytoplasm is the gooey, semi-transparent fluid in which the other organelles are suspended. Processes such as breaking down food molecules into smaller molecules happens in the cytoplasm. The cytoplasm consists mostly of water, dissolved ions, small molecules, and large water-soluble molecules such as proteins. It also has a high concentration of potassium ions and a low concentration of sodium ions, which help in controlling the amount of water in the cell. If you make a wet mount of a leaf from the pondweed *Elodea*, you may be able to see the cell's organelles, especially the chloroplasts, moving as the cytoplasm flows through the cell (Raven *et al.* 2005).

Nucleus

In eukaryote cells the nucleus is surrounded by a double membrane called the nuclear envelope. This separates the genetic material from the rest of the cell. The nuclear envelope has a large number of holes or 'pores' that allow the cell to move molecules across the nuclear envelope and in and out of the nucleus. These molecules include RNA and ribosomes moving from nucleus to the cytoplasm, and proteins, carbohydrates, and lipids moving into the nucleus.

You could say that these molecules provide a means of communication between the cytoplasm and the nucleus. The nucleus performs two important functions: it controls the activity of the cell by determining what proteins are produced and when they are produced by the cell; and it also stores the genetic information of the cell which is then passed on to daughter cells during cell division.

Vacuole

The vacuoles of most plant cells are large, filling much of the space inside the cell wall and pushing the cytoplasm out to the periphery of the cell. A single, large vacuole is found in the cytoplasm of most mature plant cells. (Animal cells have much smaller, multiple vacuoles.) The vacuole is bounded by a membrane that keeps the vacuole's contents separate from the cytoplasm. Young plant cells typically contain a number of small independent vacuoles. These merge together to form one large vacuole as the cell matures, so that in a fully mature cell the vacuole may occupy more than 90 per cent of the cell volume.

Vacuoles typically contain salts, sugars, and sometimes proteins, and have a variety of secretory, excretory, and storage functions. They are usually slightly acidic. Some of them, like those in grapefruit and lemons are very acidic – which is where the sour taste of these fruit comes from (Raven *et al.* 2005). The vacuole may also be involved in the breakdown of old organelles or macromolecules and then recycling the products back to the rest of the cell. For example, an entiremitochondrion may be engulfed by the vacuole, so that it is surrounded by a 'bubble' of the vacuole membrane. This 'bubble' is then pinched off, forming a vesicle suspended in the vacuole. After the mitochondrion has been digested the vesicle disappears.

ENDOPLASMIC RETICULUM

The endoplasmic reticulum (often abbreviated to ER) is an extensive network of membranous sack-like folds, tubes and vesicles, held together by the cytoskeleton. This network is continuous with the nuclear membrane. Plant cells contain both rough and smooth ER. The relative proportions of rough and smooth endoplasmic reticulum in a cell can change quickly, depending on the cell's metabolic needs. The endoplasmic reticulum provides a site for the synthesis of lipids and proteins and for the storage and transport of those molecules.

Rough Endoplasmic Reticulum

The surface of the rough endoplasmic reticulum is studded withribosomes, giving it a "rough" appearance (Campbell and Reece, 2005). This means that the rough ER is a site of protein synthesis. Once produced, the proteins are packaged into vesicles and transported to the Golgi apparatus, where they are modified and packaged for export from the cell or transport to where they will be used.

Smooth Endoplasmic Reticulum

Smooth endoplasmic reticulum is also made up of tubules and vesicles that branch forming a network attached to the nuclear membrane. The smooth endoplasmic reticulum provides an increased surface area for the action or storage of key enzymes and the products of these enzymes. The smooth endoplasmic reticulum also produces lipids, such as the phospholipids that make up the cell membrane - in fact, it is the source of new cell membrane material. It also metabolises carbohydrates breaking them down into glucose providing energy for the cell.

The smooth ER is also involved in the breakdown and detoxification of drugs and chemicals, *e.g.*, in human liver cells smooth ER is involved in the metabolism of ethanol. Long-term use of alcohol leads to a increased amount of smooth ER in the cells, which increases the rate of alcohol metabolism. This explains why people who regularly drink large amounts of alcohol often have a high tolerance for alcohol.

Golgi Apparatus

The Golgi apparatus is a collective term for Golgi bodies: flattened, membrane-enclosed compartments arranged into a stack of five or six pancake-like structures, often with globs or cisternae at the ends. The Golgi bodies are processing and sorting centres for cellular materials such as proteins, enzymes and lipids. Cellular materials are transported from production areas such as the endoplasmic reticulum to the Golgi bodies and, after sorting, to cisternae at the ends of the Golgi bodies. At the cisternae the molecules are enclosed in a phospholipid membrane. This membrane is then pinched off to form a vesicle, which transports the molecules to their destination and fuses with a membrane there to release its contents. In plants, cellulose sub-units are packaged into vesicles by the Golgi bodies, which then travel to and fuse with the plasma membrane. The vesicles discharge their contents to the outside of the cell and the cellulose is incorporated in the cell wall (Raven *et al.* 2005).

Mitochondria

Mitochondria—the cell's 'power plants' - generate most of the cell's supply of the energy-carrying molecule adenosine triphosphate (ATP). MItochondria are also involved in a range of other processes: signalling, cellular differentiation, cell death, control of the cell cycle, and cell growth. They evolved as the result of endosymbiosis from ancient bacteria, which were engulfed by the ancestors of eukaryote cells more than a billion years ago.

Mitochondria are much smaller than plastids - around 0.5μm in diameter, *i.e.*, too small to observe with a light microscope. The number of mitochondria in a cell varies widely: from one to several thousand, depending on organism and tissue type. Mitochondria replicate (by binary fission) mainly in response to the energy needs of the cell. In other words, their growth and division is not linked to the cell cycle. When the energy needs of a cell are high, mitochondria grow and divide. When energy use is low, mitochondria are destroyed or become inactive.

Mitochondria Structure

A mitochondrion is composed of inner and outer membranes separated by a region called the intermembrane space. The inner membrane is highly folded, forming projections called cristae.

The membrane folds greatly increase the surface area of the inner membrane, which is studded with the enzymes involved in making ATP. The area inside the inner membrane is called the matrix. The matrix contains enzymes that produce molecules used in the production of ATP on the inner membrane. The mitochondrial matrix also contains the mitochondrion's DNA and ribosomes.

Mitochondria and Aerobic Respiration

The primary role of the mitochondrion is the production of ATP, used to power cellular processes. Glycolysis in the cytoplasm produces pyruvate and NADH. Once in the mitochonrion, pyruvate is oxidised in the Krebs cycle to produce CO_2, a small amount of ATP, and hydrogen ions (H^+). These then enter the Electron Transport Chain, which generates a large amount of ATP, and water (through the combination of the H^+ ions with oxygen). In other words, if oxygen is not available other less efficient anaerobic processes, such as fermentation, must be used to generate ATP.

Aerobic respiration has a much higher yield of ATP than anaerobic respiration (Rich, 2003). For example, one molecule of glucose will produce between 32 to 36 molecules of ATP under aerobic conditions; compared to 2 molecules of ATP under anaerobic conditions.

Plastids

Like vacuoles and cell walls, plastids are characteristic of plant cells. Plastids are responsible for photosynthesis, storage of products like starch and for the synthesis of many classes of molecules, such as fatty acids and terpenes, which are needed as cellular building blocks and for normal functions of the plant. Like mitochondria, plastids originated through endosymbiosis. In plants, plastids may differentiate into several forms, depending upon which function they need to play in the cell.

Chloroplasts

Chloroplasts are give plants their characteristic green colour and are the site of photosynthesis. Like a mitochondrion, a chloroplast has an outer and inner membrane, with an intermembrane space between them. Within the inner membrane lies the stroma, which contains the thylakoid membranes—the site of photosynthesis. The thylakoid membranes form a series of flattened, fluid-filled, interconnected tubules that are often stacked on top of each other to form a structure called a granum. The photosynthetic pigments, such as chlorophyll, are located in the thylakoid membranes.

Chromoplasts

Chromoplasts - 'coloured bodies' - synthesise and store pigments. They are found in coloured organs of plants such as fruit and flower petals, giving them their distinctive colours. When fruit such as tomatoes ripen, the green chloroplasts to are converted to chromoplasts that mostly contain red carotenoids.

Amyloplasts

During the day, when plants are photosynthesising, amyloplasts store glucose and convert it into a form of starch for storage. They can also convert the starch back into sugar, when the plant needs energy.

Statoliths

Statoliths are a specialised form of amyloplast used by plants in detecting and responding to gravity. Statoliths are denser than the cytoplasm and tend to move towards the bottom of the cell, indicating which direction is 'down'. They are found mainly in root tissues.

Plant Tissues

In plants, the processes of differentiation and specialisation allow groups of cells that are structurally or functionally distinct to form tissues that have specific functions. Tissues composed of only one type of cell are called simple tissues, while those made up of two or more types of cells are called complex tissues. Parenchyma and collenchyma are simple tissues; xylem, phloem, and epidermis are complex tissues.

Tissues can also be grouped into tissue systems, and in plants these are:

- Ground tissue,
- Vascular tissue, and
- Dermal tissue.

Ground tissue is made up of parenchyma, collenchyma and sclerenchyma. Vascular tissue system consists of the two conducting tissues – xylem and phloem. Dermal tissue comprises the epidermis: pavement cells, guard cells, and trichomes. (In older woody plants the epidermis is replaced by the periderm, or bark.) (Raven *et al.* 2005).

7

Techniques for Cloning Plants

To the average consumer, plant propagation might simply mean growing a new plant from a seed. Experienced gardeners and farmers, however, know that they are numerous ways to propagate, both sexual and asexual. The asexual methods usually create an exact genetic duplicate, or clone of the parent plant. This is usually highly desirable when a gardener finds a plant with superior genetic traits. Plant cloning is almost as old as agriculture itself, and some species of trees originated as plants that were cloned on a mass scale. For example, all Bartlett pear trees have their ancestors in a tree that was first cloned in 1770. Here are the most common methods used by gardeners for cloning plants.

Layering - Layering is a method of cloning plants where roots are encouraged to actually form on the stem before it is removed from the parent plant. Some plants actually use this technique to propagate themselves in nature. In simple layering, the gardener simply bends a low growing stem into the medium, keeping it in place with a stake if necessary. The stem should form roots and start growing as a new plant. In serpentine layering, a very long stem is buried into the medium at intervals. The buried sections will form roots and create several new plants.

Grafting - Grafting is a method of plant cloning where one plant fuses with another plant. In agriculture, it is usually done with trees. In this technique, the bottom section with the plant roots is referred to as the stock, and the plant that is used for the stems and leaves is called the scion. Many farmers do this when they want to increase the odds of a plant's survival by grafting its top section to another plant with stronger roots. It is also sometimes done to save a tree that may have suffered a damaged trunk or roots.

Cutting - Cutting is the most commonly used technique for cloning plants in hydroponics. The gardener simply cuts off a part of the plant, usually a stem or leaf, and plants it into the growing medium. Since this technique requires one to actually detach a part of the plant from the root, and therefore cut off its primary source of moisture, it is imperative to keep a well humidified grow room while cloning plants through cutting. Many gardeners even choose to operate "mist sprays" on their planted cut stems during the one or two weeks that it takes for them to form roots. Micropropagation - Micropropigation is a very new, very advanced, and very expensive form of plant cloning. In this technique, lab technicians actually make clones of plants using the actual plant tissue. Since this has to be performed in a sterile laboratory environment, and is therefore very costly, it is only performed on plants that are very difficult to propagate asexually. Many believe that Micropropagation may be a way to the future of propagation, as it may have ability to create disease-free plants, as well as plants with other desirable traits, such as those that bigger yields or tastier vegetables.

TYPES OF CUTTINGS FOR CLONING PLANTS

Many resources on cloning plants sometimes refer to "plant cuttings," but don't explain the many ways that this can be done. There is real single way to cut plants. In fact, since virtually all of the cells in a plant can create every part that a plant needs to survive, a clone can be made from almost any of the plant's vegetation. Here are the four kinds of cuttings.

Stem Cutting

Just as its name implies, stem cutting involves taking a stem from the plant to make a clone. Stem cutting is perhaps the most popular cutting method for cloning plants in hydroponics. A stem cutting should be a couple inches in length and have a few healthy, large leaves to maximize photosynthesis. Make certain to keep the leaves of stem moist as it starts to take root.

Leaf Cutting

In this type of cutting for cloning plants, the leaf blade itself is used to create a new plant. Usually, the cut leaf doesn't become a part of the new plant.

If the plant you are growing has particularly thick leaves, you can cut open the veins of the leaf and plant it flat into the growing medium, making sure to keep the cut leaf exposed to light and moisture. After a short period of time, new plants may begin to form where the leaf was cut open. The cut leaf will eventually dry up and rot away, even as the young plants thrive.

Another technique sometimes used is to take older, larger leaves and cut them into triangles, making sure each piece contains a large vein. The edges of the leaf, which usually contains no veins, should be thrown away. The triangles are then planted in the medium with the pointy side down. A new plant can then develop from the vein.Be aware, however, that leaf cutting is very delicate, and only works with plants that are easily cloned.

Leaf-Bud Cutting

This type of cutting is made up of a leaf blade, petiole, and a short piece of the stem with an attached bud. Leaf bud cuttings are best of species that are able to create roots but not shoots from cut leaves. This method works best when you have a healthy plant, but have very little cloning material to work with because it maximizes the propagating material.

Root Cutting

Propagation by cutting a parent's plant's roots is very simple, because there is less concern of the roots taking hold. However, it is not as popular because many gardeners don't want to harm the roots of their plants. Obviously, the thicker and healthier the root, the greater the chance of success. Roots should be two or three inches in length, and planted right side up in the growing medium. Many growers choose to cut the side that is supposed to face upwards with a straight cut and the side that is supposed to face downwards with a diagonal cut so that they do not accidentally plant it wrong.No matter what kind of cutting you choose for cloning plants, make certain to use a quality rooting gel to stimulate the formation of new roots.

METHOD OF DNA CLONING AND MANIPULATE DNA

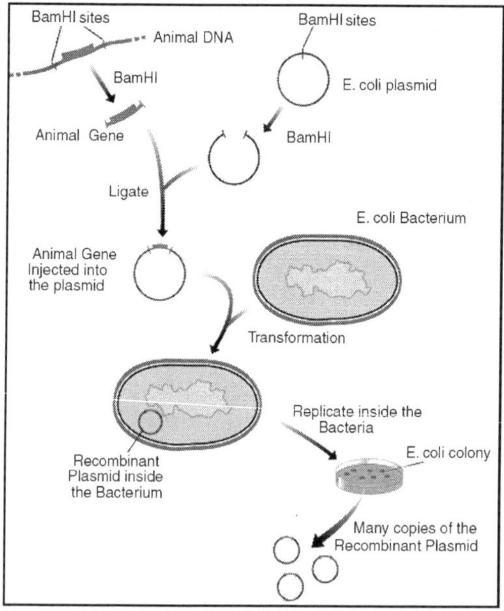

Fig. An Outline of Gene Cloning.

DNA cloning a method to manipulate DNA *in vitro* and *in vivo* or is a logical extension of the ability to manipulate DNA molecules with restriction endonucleases and ligases. Imagine that an plant gene has been obtained as a single restriction fragment after digestion of a larger molecule with the restriction enzyme *Bam*HI, which leaves 5-GATC-3 sticky ends. Imagine also that a small *E. coli* plasmid has been purified and treated with *Bam*HI, which cuts the plasmid in a single position. The circular plasmid has therefore been converted into a linear molecule, again with 5-GATC-3 sticky ends. Mix the two DNA molecules together and add DNA ligase. Various recombinant ligation products will be obtained, one of which comprises the circularized plasmid with the animal gene inserted into the position originally taken by the *Bam*HI restriction site. If the recombinant plasmid is now re-introduced into *E. coli*, and the inserted gene has not disrupted its replicative ability, then the plasmid plus inserted gene will be replicated and copies passed to the daughter bacteria after cell division. More rounds of plasmid replication and cell division will result in a colony of recombinant *E. coli* bacteria, each bacterium containing multiple copies of the animal gene. This series of events, constitutes the process called DNA or gene cloning.

VECTOR THAT WORKS WITH PLANT CELLS

Not all aspects of the genetic manipulation of plant cells are readily accomplished. Not only do plants usually have a great deal of chromosomal material and grow relatively slowly as compared with single cells grown in the laboratory, but few cloning vectors can successfully function in plant cells. While researchers working with animal cells can choose among a wide variety of cloning vectors to find just the right one, plant cell researchers are currently limited to just a few basic types of vectors.

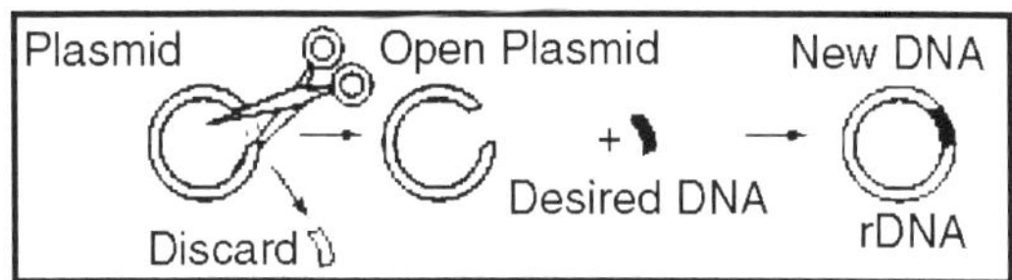

Fig. Insertion of Foreign DNA in Plasmid

Perhaps the most commonly used plant-cloning vector is the "Ti" plasmid, or tumor-inducing plasmid. This plasmid is found in cells of the bacterium known as Agrobacterium tumefaciens, which normally lives in soil. The bacterium has the ability to infect plants and cause a crown gall, or tumorous lump, to form at the site of infection. The tumor-inducing capacity of this bacterium results from the presence of the Ti plasmid.

The Ti plasmid itself, a large, circular, double-stranded DNA molecule, can replicate independently of the A. tumefaciens genome. When these bacteria infect a plant cell, a 30,000 base-pair segment of the Ti plasmid - called T DNA - separates from the plasmid and incorporates into the host cell genome. This aspect of Ti plasmid function has made it useful as a plant cloning vector.

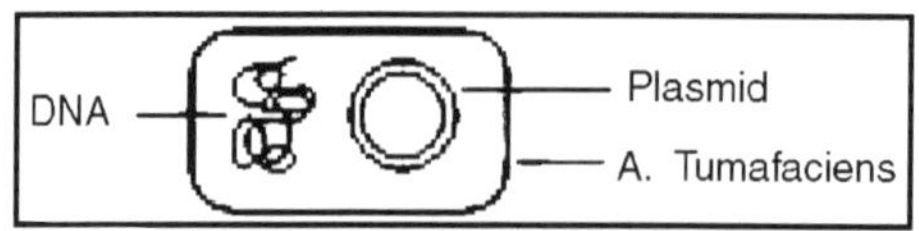

Fig. A. Tumefiecian

The Ti plasmid can be used to shuttle exogenous genes into host plant cells.

This type of gene transfer requires two steps:

1. The endogenous, tumor-causing genes of the T DNA must be inactivated
2. Foreign genes must be inserted into the same region of the Ti plasmid.

The resulting recombinant plasmid, carrying up to approximately 40,000 base pairs of inserted DNA and including the appropriate plant regulatory sequences, can then be placed back into the A. tumefaciens cell. That cell can be introduced into plant cell protoplasts either by the process of infection or by direct insertion. Once in the protoplast, the foreign DNA, consisting of both T DNA and the inserted gene, incorporates into the host plant genome. The engineered protoplast - containing the recombinant T DNA - regenerates into a whole plant, each cell of which contains the inserted gene. Once a plant incorporates the T DNA with its inserted gene, it passes it on to future generations of the plant with a normal pattern of Mendelian inheritance.

One of the earliest experiments that involved the transport of a foreign gene by the Ti plasmid involved the insertion of a gene isolated from a bean plant into a host tobacco plant. Although this experiment served no commercially useful purpose, it successfully established the ability of the Ti plasmid to carry genes into plant host cells, where they could be incorporated and expressed.

A. Limition of A. Tumifecian

The fact that only certain types of plants were naturally susceptible to infection with the host bacterial organism initially limited the usefulness of the Ti plasmid as a cloning vector. In nature, A. tumefaciens infects only dicotyledons or "dicots" - plants with two embryonic leaves. Dicotyledenous plants, divided into approximately 170,000 different species, include such plants as roses, apples, soybeans, potatoes, pears, and tobacco. Unfortunately, many important crop plants, including corn, rice, and wheat, are monocotyledons - plants with only one embryonic leaf - and thus could not be easily transfected using this bacterium.

Overcoming the Limition

Research efforts in the past few years have reduced the limitations of A. tumefaciens. Scientists discovered that by using the processes of microinjection, electroporation, and particle bombardment, naked DNA molecules can be introduced into plant cell types that are not susceptible to A. tumefaciens transfection. Microinjection involves the direct injection of material into a host cell using a finely drawn micropipette needle. Electroporation uses brief pulses of high voltage electricity to induce the formation of transient pores in the membrane of the host cell. Such pores appear to act as passageways through which the naked DNA can enter the host cell.

Particle bombardment actually shoots DNA-coated microscopic pellets through a plant cell wall. These developments, important in the commercial application of plant genetic engineering, render the valuable food crops of corn, rice, and wheat susceptible to a variety of manipulations by the techniques of recombinant DNA and biotechnology.

CLONING UNICELLULAR ORGANISMS

Cloning a cell means to derive a population of cells from a single cell. In the case of unicellular organisms such as bacteria and yeast, this process is remarkably simple and essentially only requires the inoculation of the appropriate medium. However, in the case of cell cultures from multi-cellular organisms, cell cloning is an arduous task as these cells will not readily grow in standard media.

A useful tissue culture technique used to clone distinct lineages of cell lines involves the use of cloning rings (cylinders). According to this technique, a single-cell suspension of cells that have been exposed to a mutagenic agent or drug used to driveselection is plated at high dilution to create isolated colonies; each arising from a single and potentially clonal distinct cell. At an early growth stage when colonies consist of only a few of cells, sterile polystyrene rings (cloning rings), which have been dipped in grease are placed over an individual colony and a small amount of trypsin is added. Cloned cells are collected from inside the ring and transferred to a new vessel for further growth.

Cloning Stem Cells

Somatic-cell nuclear transfer, known as SCNT, can also be used to create embryos for research or therapeutic purposes. The most likely purpose for this is to produce embryos for use in stem cell research. This process is also called "research cloning" or "therapeutic cloning." The goal is not to create cloned human beings (called "reproductive cloning"), but rather to harvest stem cells that can be used to study human development and to potentially treat disease. While a clonal human blastocyst has been created, stem cell lines are yet to be isolated from a clonal source. Therapeutic cloning is achieved by creating embryonic stem cells in the hopes of treating diseases such as diabetes and Alzheimer's. The process begins by taking out the nucleus (containing the DNA) from an egg cell and putting in it a nucleus from the adult cell to be cloned. In the case of someone with Alzheimer's disease, the nucleus from a skin cell of that patient is placed into an empty egg. The reprogrammed cell begins to develop into an embryo because the egg reacts with the transferred nucleus. The embryo will become genetically identical to the patient. The embryo will then form a blastocyst which has the potential to form/become any cell in the body.

The reason why SCNT is used for cloning is because somatic cells can be easily acquired and cultured in the lab. This process can either add or delete specific genomes of farm animals. A key point to remember is that cloning is achieved when the oocyte maintains its normal functions and instead of using sperm and egg genomes to replicate, the oocyte is inserted into the donor's somatic cell nucleus. The oocyte will react on the somatic cell nucleus, the same way it would on sperm cells.

The process of cloning a particular farm animal using SCNT is relatively the same for all animals. The first step is to collect the somatic cells from the animal that will be cloned. The somatic cells could be used immediately or stored in the laboratory for later use. The hardest part of SCNT is removing maternal DNA from an oocyte at metaphase II. Once this has been done, the somatic nucleus can be inserted into an egg cytoplasm. This creates a one-cell embryo. The grouped somatic cell and egg cytoplasm are then introduced to an electrical current. This energy will hopefully allow the cloned embryo to begin development. The successfully developed embryos are then placed in surrogate recipients, such as a cow or sheep in the case of farm animals.

SCNT is seen as a good method for producing agriculture animals for food consumption. It successfully cloned sheep, cattle, goats, and pigs. Another benefit is SCNT is seen as a solution to clone endangered species that are on the verge of going extinct. However, stresses placed on both the egg cell and the introduced nucleus are enormous, leading to a high loss in resulting cells. For example, the cloned sheep Dolly was born after 277 eggs were used for SCNT, which created 29 viable embryos. Only three of these embryos survived until birth, and only one survived to adulthood. As the procedure currently cannot be automated, and has to be performed manually under amicroscope, SCNT is very resource intensive. The biochemistry involved in reprogramming the differentiated somatic cell nucleus and activating the recipient egg is also far from being well-understood.

In SCNT, not all of the donor cell's genetic information is transferred, as the donor cell's mitochondria that contain their own mitochondrial DNA are left behind. The resulting hybrid cells retain those mitochondrial structures which originally belonged to the egg. As a consequence, clones such as Dolly that are born from SCNT are not perfect copies of the donor of the nucleus.

PLANT GENE CLONING

Plant gene cloning is currently limited to simply inherited single gene traits. The transfer of genes to plants is limited to single genes or, at most, two to three genes encoding a particular, well-characterized biochemical pathway. Manipulation of multigenic traits is currently only feasible within the existing gene pool, sometimes made more efficient by markerassisted selection. When traits are regarded in the molecular genetic sense, often the coding sequence is the major consideration. Obviously, the coding sequence is of major importance as this is the template for determining the amino acid sequence of the final protein or enzyme which determines the phenotype. However, a major factor in controlling the phenotype is the presence of regulatory sequences surrounding the coding sequence in the form of upstream (promoters and enhancers) and downstream (terminator) regulatory sequences and, in many eukaryotic genes, introns. When traits are being manipulated via genetic engineering, the role of these regulatory sequences cannot be trivialized. Put simply, these sequences control such things as the level of gene expression, tissue/organ specificity, developmental expression and response to particular abiotic/biotic stimuli.

The regulatory sequences that have been subjected to greatest attention in plants are the upstream promoters. For proper expression of a seed storage protein, for example, the gene of interest must be under the control of a promoter which expresses in developing endosperm (monocots) or cotyledons (dicots). Expression of other genes may be required only in anthers, root hairs or photosynthetic tissues, or in response to a wound such as that caused by a chewing insect or penetration of a fungal infection structure. In the earliest cases of transgenic plants, introduced genes were placed under the control of constitutive promoters of bacterial (octopine or nopaline synthetase) or viral (cauliflower mosaic virus (CaMV) 35S) origin. These promoters have been found to give adequate expression of selectable marker and reporter genes in dicotyledonous species.

However, expression is commonly not of a sufficient level in monocots, especially cereals, for good selection with antibiotic or herbicide resistance genes. And, as already discussed, such a constitutive pattern of gene expression is commonly not desirable for most traits. For cereals, a number of other promoters have been developed to achieve high-level constitutive expression of selectable markers or reporter genes. These include sequences cloned directly from cereal tissues such as the ubiquitin and rice actin 1 promoters and synthetic promoters involving various spliced sequences from a range of sources such as the Emu promoter. However, to properly control gene expression such that the introduced character is most efficiently targeted without other pleiotropic effects, it is generally found that regulatory sequences of plant origin are desirable. Further to this, expression is usually best controlled when a cereal promoter is used in a cereal or a dicot promoter in a dicot. The most widely used means of cloning such regulatory sequences is by using cDNA libraries from specific tissue types and looking for abundant signals. Naturally, cDNA libraries will not contain the regulatory sequences, hence the abundant cDNA clones must then be used to screen a genomic library so that the regulatory sequences can be identified and cloned.

METHODS FOR CLONING PLANT TRAITS

Cloning Based on Knowledge of Protein Structure/sequence

Many of the first plant traits to be cloned were those in which some degree of biochemical understanding preceded any work with DNA. The understanding may have come from plant systems or from outside the plant kingdom. Most of the cloned genes in these cases were enzymes which were part of a well-characterized biochemical pathway or had a protein structure which was otherwise well known, such as seed storage proteins.

Cloning genes based on a knowledge of the protein sequence can be achieved by working backwards from the gene product (amino acid sequence) to the coding sequence or the DNA template. Such an approach has been used for cloning many seed storage proteins, such as the 2S sulphur- rich proteins from Brazil nut. Genes of plant origin have been used in transgenic plants to improve resistance to insect pests. One of the first to be characterized was the cowpea trypsin inhibitor, a protease inhibitor which acts as an antifeedant to a range of insect pests.

The protein is naturally produced in seeds of some genotypes of cowpea (*Vigna unguiculata*), which was shown to confer lepidopteran insect resistance when expressed constitutively in tobacco. Another such report from the same group at Durham University described the use of a snowdrop lectin gene to confer aphid resistance in transgenic tobacco, the first report of genetically engineered resistance to a sucking insect in plants. The lectin from snowdrop (*Galanthus nivalis*), known as GNA, has been demonstrated to be effective against two major rice pests, the brown planthopper and the green leafhopper, as well as the peach potato aphid. When expressed at 0.1% of total leaf protein in tobacco, GNA is antimetabolic to homopteran pests. These levels of protein were accumulated in leaf tissues when placed under the control of the CaMV 35S promoter. However, Hilder *et al.* (1995) postulate that better control would be achieved with the use of a phloem-specific promoter, such as the sucrose synthase 1 promoter from rice.

Another significant group of genes which have been cloned based on prior knowledge of the end product are the seed storage genes, including both seed storage proteins and lipids. In both cases there was considerable biochemical knowledge of the pathways involved in biosynthesis and storage. For seed storage proteins, the genes can be cloned based on the knowledge of the amino acid sequence of the mature protein and targeting signal or transit peptide. Molecular genetic manipulation of vegetable oils is a major area of research in the developed world. While approximately 90% of vegetable oils are used for human consumption, there are emerging markets for industrial applications including lubricants, paints, cosmetics, plasticizers and soaps.

With increasing emphasis on quality and dietary considerations, there is a desire in developed countries to substantially alter the types and balances of fatty acids found in the major oil crops. Six crop species (soybean, oil palm, rapeseed/canola, sunflower, cottonseed and peanut) account for 84% of the world's vegetable oil production, with the major fatty acids produced being palmitic, stearic, linoleic, linolenic and oleic acids. However, almost 200 different fatty acids are produced by plants, mostly occurring in nondomesticated plant species. Hence, genetic manipulation of the major crop species, either via conventional breeding or genetic engineering, has the potential to modify the fatty acid types produced and stored by seeds such as soybean and canola.

The biochemistry of fatty acid biosynthesis is well established for many of the major pathways, with enzymes identified which control the chain length of fatty acids, and the degree of saturation, or number of double bonds in the fatty acid chain. The oils stored in seeds have been altered by downregulation of particular enzymes using antisense RNA techniques. The oleic acid content of rapeseed oil has been increased to 83% (from 62% in conventionally bred cultivars) by downregulating the enzyme D12-oleate desaturase using antisense RNA. A similar result has been achieved using co-suppression, which has produced rapeseed oil with 87% oleic acid content. Genetic engineering approaches have also been used to introduce totally novel fatty acids to major oilseed crops. Petroselinic acid, an isomer of oleic acid, is found in abundance in the spice plant, coriander (*Coriandrum sativum*). This oil has uses in cosmetics and pharmaceuticals, and its oxidation leads to the formation of lauric acid, used in detergents, and adipic acid, which is used in nylon production.

Hence this is a potentially valuable oil which may enable soybean and canola crops – i.e. renewable resources – to replace some of the nonrenewable carbon fuels which are currently used for many industrial polymer and lubrication applications. A cDNA clone from coriander encoding an acyl-ACP desaturase has been genetically engineered into tobacco, resulting in the accumulation of petroselinic acid.

Oil crops such as rapeseed and soybean could be genetically modified to produce this fatty acid on a commercial scale. Hence, here is an example whereby a minor crop plant has become a donor of genetic diversity which will enable the commercial production of a very useful array of products for food and industrial purposes.

Cloning Genes Known Only by Phenotype

However, a more challenging task has been to move beyond proteins and pathways about which sufficient biochemistry is known into cloning genes known by phenotype only. By far the majority of genes fit into this category, which includes many of the most important plant traits such as disease resistance, adaptation to abiotic stresses, phenology and flowering behaviour and fertility. Only in the past few years have genes for traits known only by phenotype been cloned. One of the most interesting areas of plant biology is the understanding of plant–pathogen interactions at the molecular level. The major methods used for cloning genes known only by phenotype are:

- Transposon tagging.
- Map-based or positional cloning.
- T-DNA tagging.

There have been a number of plant genes cloned using all these methods or modifications of them. There are also examples where a combination of methods has been used to expedite the rate of identification of the unknown sequence.

Transposon Tagging

The proposal that specific pieces of DNA are able to autonomously excise from a particular locus and insert, theoretically at random, into another linked or nonlinked locus was first proposed by Barbara McClintock (1949, 1950). Transposable elements are now accepted as ubiquitous, and are particularly well characterized in organisms such as *Escherichia coli*, *Drosophila* and maize. In two plant species, maize and the snapdragon (*Antirrhinum majus*), transposable elements are sufficiently well characterized to be used for mutagenesis and gene tagging.

The best understood system is the maize *Ac*/*Ds* (*Activator*/*Dissociation*) family of elements. The behaviour of the *Ac* element is such that it can excise autonomously from its locus, under the control of its own transposase, and insert into another chromosomal position. Where the *Ac* element is in or near a gene, that gene is inactivated (i.e. not expressed), and will appear phenotypically as a recessive mutation. When *Ac* transposes (excises and reinserts at another locus), the gene is commonly returned to normal function, and hence is termed a revertant.

Should this *Ac* element then insert into another gene (transcribed or regulatory sequence), then a new recessive mutation will manifest. Transposition can occur either germinally, which results in an altered phenotype for the whole plant, or somatically, which results in a mosaic mutation, and can often be detected as a sectoring of plant or seed colour. The *Ds* element is actually a group of elements derived from defective *Ac* elements, with a common feature being their lack of effective *Ac* transposase.

Table. Plant Transposable Elements which have been used for Transposon-Taggine

Element	*Species*	*Type of Element*	*Size*
Ac	Maize	Autonomous	4.6
Ds	Maize	Nonautonomous	0.5–30]
Tam1	Snapdragon	Autonomous	15
Tam2	Snapdragon	Nonautonomous	5
En/Spm	Maize	Autonomous	8.3

Hence these elements are nonautonomous, and will only transpose in the presence of an *Ac* element capable of producing transposase. Equipped with sequence and map information of an *Ac* element in maize, geneticists have been able to clone important maize genes using the transposon-tagging approach. When a gene is inactivated by an *Ac* element, proof of which can be demonstrated by a low frequency of trait reversion, the *Ac* element can be used as a probe for 'fishing-out' the tagged gene.

A tagged gene can be cloned by performing Southern blot analysis with the *Ac* element used as probe, and any polymorphism represents an excision/ insertion event. The technique is described more fully in the review by Walbot (1992). The *Ac* system has proven extremely useful for cloning maize genes for traits such as plant pigment genes (*bronze*; the anthocyanin pathway), starch biosynthesis (*opaque2*), and abscisic acid (AbA) insensitivity (McCarty *et al.*, 1989). Other maize transposable elements such as *Mutator* (*Mu*) and the *En/ Spm* system have been used to clone maize genes, as has the *Tam* system in snapdragon. As a result, the search began for transposable element systems in other species so that a similar approach could be taken.

However, it was soon demonstrated that the well-characterized *Ac/Ds* system could be genetically engineered into other plant species including tobacco, *Arabidopsis*, tomato and petunia. It was subsequently shown that the *Ac* element acted in a similar manner in these heterologous species, undergoing germinal and somatic excision and insertion, and maintaining the propensity to reinsert at a linked site (usually within 4 cM). As a result, the frequency of mutation in the surrounding area approaches 1024. It was not until 1993 that a plant gene was cloned using the heterologous system, that of the *Ph6* gene in petunia, which is involved in floral pigmentation.

Using the transposon-tagging approach, a number of plant disease resistance genes (R genes) have now been cloned. The first of these was the tomato *Cf-9* gene, which confers resistance to race 9 of the fungal pathogen *Cladosporium fulvum*, causal agent of leaf mould. The gene was tagged using a modified *Ac/Ds* system. The system relied on the use of a stabilized *Ac* element (s*Ac*), which was effectively a disabled *Ac* element that nevertheless produced *Ac* transposase and hence enables *Ds* transposition. The two elements are maintained in separate lines, and hence are both stable until controlled crossing is performed. This gives a significant advantage in that *Ds* insertion mutants can be maintained indefinitely in the homozygous or hemizygous form, and revertants can be isolated by crossing lines with s*Ac* tomatoes.

The transposon-tagging approach used to isolate the *Cf-9* gene, which was known by phenotype only, did have considerable advantages in that much was known of the pathogen. The hypersensitive response elicited by the virulent strain fitted into the gene-for-gene hypersensitive response proposed by Flor (1971), with a dominant avirulence gene (*Avr9*) interacting with the *Cf-9* gene to form an incompatible or hypersensitive response. *Avr9* had previously been cloned and fully characterized, and was known to specify a 28 amino acid peptide which alone could elicit a necrotic response in resistant tomatoes.

Crosses were made to develop a transgenic tomato line homozygous for *Cf-9*, but heterozygous for *Ds*. Crossing was performed as outlined in Fig.. As can be seen, most seedlings died from systemic hypersensitive response, with *Avr9* and *Cf-9* present in all cells. The only survivors would be individuals where *Ds* had inserted into the *Cf-9* gene. Where this happened in plants carrying s*Ac*, the seedlings were variegated for the hypersensitive response (somatic excision), and in the absence of s*Ac*, the survivors were stable (germinal excision).

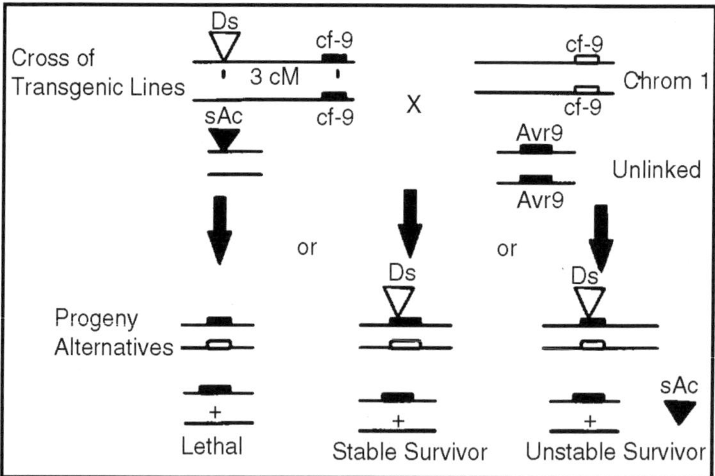

Fig. Graphical Representation of the Crosing Strategy of Transgenic Ac/Ds Tomato Lines.

This work was a major undertaking, involving the germination of approximately 160 000 progeny, yielding 118 survivors, representing 63 independent mutations. After considerable effort to characterize these, the

Cf-9 gene was cloned and characterized, with an open reading frame encoding an 863 amino acid protein.

Map-based Cloning

The previously described transposon-tagging scheme utilized to clone *Cf- 9* had an element of the map-based approach by utilizing the phenomenon of *Ac/Ds* transposition to linked sites, with a *Ds* element within 3 cM of the locus of interest, which greatly enhanced the likelihood of insertional mutagenesis. However, map-based cloning requires a much tighter genetic linkage (in centimorgans or recombination fraction), which in turn needs to be converted into a physical distance (kbp of DNA sequence).

The map-based, or positional cloning concept relies on building a well-saturated genetic map of DNA sequences of the organism in question. The genetic map is built up in a polymorphic mapping population, commonly an F2, backcross (BC) or recombinant inbred, with markers such as RFLPs (restriction fragment length polymorphisms). The first RFLP map to be made was of human. Based on such maps, important human genes responsible for disorders such as cystic fibrosis and muscular dystrophy have been cloned in this manner.

Genetic maps of plant species have been made with DNA markers in virtually all important plant species, particularly the major cereals, grain legumes, oilseeds and important vegetable crops including tomato and potato. The RFLP marker system is based on Southern blot analysis, while many of the newer technologies use the polymerase chain reaction (PCR). These include random amplified polymorphic DNA (RAPDs), amplified fragment length polymorphisms (AFLPs), and a number of different marker systems based on simple sequence repeats (SSRs), which may use PCR, such as inter simple sequence repeat (ISSR)-PCR, or Southern analysis. The applications of DNA markers to plant genetic analysis and breeding are many, and the subject has been reviewed by various authors.

The population for saturation mapping should be segregating in a Mendelian manner for the trait of interest. This requires a lot of tedious but demanding work, and it may be that particular populations such as near isogenic lines, or DNA pooling strategies, such as bulked segregant analysis, will allow the mapping effort to be more targeted to the region of interest, hence improving the rate of map saturation.

It is desirable to obtain extremely close genetic linkage to the locus of interest (<1 cM), with flanking markers if possible. It can be of considerable advantage to identify a marker which co-segregates (say within 0.1 cM) with the trait. The next step is to convert the genetic map into a physical map of the chromosome region, translating distances from centimorgans to nucleotides.

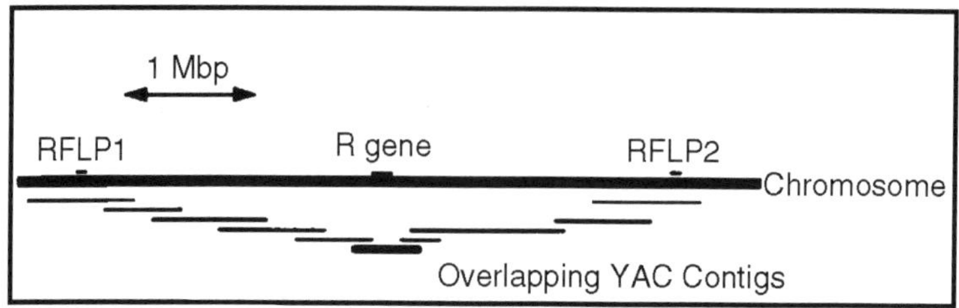

Fig. Generalized Strategy of Chromosome Walking Used in Map-based gene Cloning.

The Gene of Interest (in this illustration,R) is Flanked by Markers such as RFLPs or SSRs.A Series of Overlapping YAC Contigs is used to cover th Chromosomal Region Containing the Gene.The YAC Which Contains the Complete Gene is Marked in Bold.This is then Subcloned or Matched to a cDNA Library to Idenfify the Gene(s) Contained the Region, and Ultimately Identify the R Gene.

Physical mapping then involves the use of YAC or BAC clones which span the region of interest. These YAC/BAC contigs will span the region between two flanking DNA marker sequences as shown in Fig., a process known as chromosome walking. In this example, two YAC/BAC clones contain the gene of interest. Methodology is being developed to enable BAC clones to be directly transformed into plants, to find which one contains the gene of interest (in the example, the *R* gene). Alternatively, the YAC/BAC must be subcloned (for example as a cDNA library), and these subclones can be put into the susceptible plant and tested for the R phenotype.

The first R gene to be cloned in plants via map-based cloning was the *Pto* gene, which confers resistance to *Pseudomonas syringae* pv. *lycopersici*, the causal agent of bacterial speck in tomato. The starting points for the work were:

- A high-density RFLP map of tomato.
- A tomato YAC library.

The *Pto* locus had been previously mapped to tomato chromosome 5 and one RFLP marker co-segregated with *Pto*. Hence this was a useful starting point to screen the YAC library. A 400 kb YAC was identified, and via inverse PCR, some end-specific probes were generated. The left arm of the YAC was 1.8 cM from *Pto*, with the right arm perfectly co-segregating. This YAC was then used to screen a 920 000 plaque leaf cDNA library, from which 30 were selected. One of these was shown to co-segregate with *Pto*. This cDNA was then transformed into a susceptible tomato cultivar under the control of the CaMV 35S promoter, where it conferred resistance. Hence the gene was identified, and could then be sequenced and characterized. The success of this approach was partly due to the small genome size of tomato (950 Mbp), and a relatively low level of repeat sequence. Such an approach is much more difficult in species with larger genomes such as maize (2300–2700 Mbp) and wheat (16 000 Mbp). One such example of this problem is the 79 000 clone maize YAC library produced by Zeneca Seeds. As reported by K. Edwards, only 15% of the YACs have unique ends, with the remainder terminating in repeat sequences, which makes chromosome walking very difficult. The synteny of genomes is a major advantage in this instance.

It is now well established that there is a high degree of synteny of genome organization, as seen in the co-linearity of maps of the most important grass genomes. This has resulted in the great interest in using this co-linearity to facilitate cloning of genes in large genomes such as maize and wheat, by starting with the smaller genomes, particularly rice, and to a lesser extent, sorghum. One particular study of maize YACs demonstrated that a single YAC with the *Adh1* sequence contained 36 different repetitive sequences, whereas these repetitive sequences were largely absent in corresponding areas of the rice and sorghum genomes, a pattern seen also for the *a1–sh2* region in all three genomes.

In a similar manner, many genes of interest for *Brassica* improvement are being identified in the smaller *Arabidopsis thaliana* genome. Paterson *et al.* (1995) have recently demonstrated the potential of this approach, by identifying quantitative trait loci (QTLs) for seed mass and phenology which were co-linear on molecular marker maps of maize, sorghum and rice. They proposed that, as the maize genome is four and six times larger than the sorghum and rice genomes, respectively, the genes of interest in maize may be cloned in sorghum/rice more expediently. It must be remembered that QTLs are subject to genotype 3 environment interaction, which will be a major complicating factor in attempts to clone and characterize a QTL.

T-DNA Tagging

When T-DNA is inserted into the plant genome, it does so via illegitimate recombination. In the same way as a transposon, the insertion of T-DNA into a gene will inactivate it, which phenotypically will appear as a mutation. The comparative advantage of such a method is that stable mutations will be conferred. This can be a disadvantage, however, as reversions, such as those made possible by the *Ac/Ds* system are not possible. It is also the case that each independent transformation event will result in just one mutation event (or more if multiple independent insertions are made), whereas the heterologous transposon approach allows multiple mutation events to take place from a single initial transformant. Hence, mutated genes with T-DNA insertions can be cloned, in much the same manner as those with transposon insertions, by using part of the T-DNA as a probe in Southern blots.

The procedure for selecting out the tagged gene is quite straightforward, particularly when the plant contains a single T-DNA copy. However, this means that a large number of independent transformants must be generated. This can be achieved with species such as *Arabidopsis*, where Feldmann (1991) demonstrated that a seed-based transformation system using *Agrobacterium* can be quite efficient. Feldmann (1991) demonstrated that by inoculating approximately 1000 seeds, approximately 300 000 selfed T2 seed can be collected. They estimated that approximately 8000 transformants could be recovered from an experiment of this scale and, using this approach, cloned a gene involved in trichome development and a dwarf gene from *Arabidopsis*.

A regulatory gene in the ethylene biosynthesis pathway of *Arabidopsis* has also been cloned using the T-DNA tagging strategy. With modification, T-DNA tagging methodology can be used to clone plant promoter sequences. Koncz *et al.* (1989) designed a vector which contained a promoterless antibiotic selection gene (aminoglycoside phosphotransferase II) and transformed this into *Nicotiana* and *Arabidopsis*. In this way, promoters could be 'trapped' as plants that were regenerated on selective antibiotic were antibiotic resistant because the T-DNA was inserted downstream of a promoter such that the gene was expressed.

PRODUCING CLONES: PLANT LIFE

Nature- has been cloning organisms for billions of years. For example, when a strawberry plant sends out a runner (a form of modified stem), a new plant grows where the runner takes root. That new plant is a clone. Similar cloning occurs in grass, potatoes and onions. People have been cloning plants in one way or another for thousands of years. For example, when you take a leaf cutting from a plant and grow it into a new plant (vegetative propagation), you are cloning the original plant because the new plant has the same genetic makeup as the donor plant. Vegetative propagation works because the end of the cutting forms a mass of non-specialized cells called a callus. With luck, the callus will grow, divide and form various specialized cells (roots, stems), eventually forming a new plant.

More recently, scientists have been able to clone plants by taking pieces of specialized roots, breaking them up into root cells and growing the root cells in a nutrient-rich culture. In culture, the specialized cells become unspecialized (dedifferentiated) into calluses. The calluses can then be stimulated with the appropriate plant hormones to grow into new plants that are identical to the original plant from which the root pieces were taken. This procedure, called tissue culture propagation, has been widely used by horticulturists to grow prized orchids and other rare flowers.

PRODUCING CLONES: ANIMAL KINGDOM

Plants are not the only organisms that can be cloned naturally. The unfer-tilized eggs of some animals (small invertebrates, worms, some species of fish, lizards and frogs) can develop into full-grown adults under certain environmental conditions -- usually a chemical stimulus of some kind. This process is called parthenogenesis, and the offspring are clones of the females that laid the eggs. Another example of natural cloning is identical twins. Although they are genetically different from their parents, identical twins are naturally occurring clones of each other.

Scientists have experimented with animal cloning, but have never been able to stimulate a specialized (differentiated) cell to produce a new organism directly. Instead, they rely on transplanting the genetic information from a specialized cell into an unfertilized egg cell whose genetic information has been destroyed or physically removed. In the 1970s, a scientist named John Gurdon successfully cloned tadpoles. He transplanted the nucleus from a specialized cell of one frog (B) into an unfertilized egg of another frog (A) in which the nucleus had been destroyed by ultraviolet light. The egg with the transplanted nucleus developed into a tadpole that was genetically identical to frog B. While Gurdon's tadpoles did not survive to grow into adult frogs, his experiment showed that the process of specialization in animal cells was reversible, and his technique of nuclear transfer paved the way for later cloning successes.

Dolly

In 1996, cloning was revolutionized when Ian Wilmut and his colleagues at the Roslin- Institute in Edinburgh, Scotland, successfully cloned a sheep named Dolly. Dolly was the first cloned mammal. Wilmut and his colleagues transplanted a nucleus from a mammary gland cell of a Finn Dorsett sheep into the enucleated egg of a Scottish blackface ewe. The nucleus-egg combination was stimulated with electricity to fuse the two and to stimulate cell division. The new cell divided and was placed in the uterus of a blackface ewe to develop. Dolly was born months later. Dolly was shown to be genetically identical to the Finn Dorsett mammary cells and not to the blackface ewe, which clearly demonstrated that she was a successful clone (it took 276 attempts before the experiment was successful). Dolly has since grown and reproduced several offspring of her own through normal sexual means. Therefore, Dolly is a viable, healthy clone. Since Dolly, several university laboratories and companies have used various modifications of the nuclear transfer technique to produce cloned mammals, including cows, pigs, monkeys, mice and Noah.

8

Composite Scaffolds for Bone Tissue Engineering

Since the late 1980s, tissue engineering (TE) as a new discipline has made rapid advances. Tissue engineering, as a potential medical treatment, holds promises of:

- Eliminating re-operations by using biological substitutes,
- Using biological substitutes to solve problems of implant rejection, transmission of diseases associated with xenografts and allografts, and shortage in organ donation,
- Providing long-term solutions in tissue repair or treatment of diseases,
- Potentially offering treatments for medical conditions that are currently untreatable.

Tissue engineering has therefore attracted great attention in science, engineering, medicine, and the society. It is now generally recognised that one of the key issues in tissue engineering is the development of suitable biodegradable materials and scaffolds for seeding cells and for the subsequent growth of tissues.

For bone tissue engineering, major issues of tissue engineering scaffolds include the use of appropriate matrix materials for scaffolds, control of porosity and pore characteristics of scaffolds, mechanical strength of scaffolds, scaffold degradation properties, and bioactivity (i.e., osteoconductivity, or osteoinductivity if possible) of scaffolds.

Biomaterials experience with biodegradable polymers in the pre-TE time has heavily guided researchers towards using medical-profession-accepted and the US Food and Drug Administration (FDA)-approved biopolymers such as poly(lactic acid) (PLA) and poly(lactic acid-co-glycolic acid) (PLGA) for constructing tissue engineering scaffolds. In recent years, other polymers have also been developed for tissue engineering applications. Despite their phenomenal developments in the past three decades, bioactive bioceramics (including glasses and glass-ceramics) have been overlooked for tissue engineering applications until recently. It has been shown that some bioactive glasses have the potential for bone tissue engineering and there are now reports on the development of tissue engineering scaffolds purely made of bioceramics.

Compared to the strengths of metals and ceramics for medical applications, the strengths of biodegradable polymers are already very low. With the introduction of pores in the polymers to form tissue engineering scaffolds, the strengths of porous structures are further decreased, as strengths of materials decrease drastically with an increase in porosity. On the other hand, polymers such as PLA and poly (PCL) are considered to be non-osteoconductive. In order to seek better scaffolds for bone tissue engineering, a composite strategy can therefore be adopted. Polymer-based scaffolds containing bioactive bioceramics can be produced in which the bioceramics can serve two purposes:

- Making the scaffolds osteoconductive,
- Reinforcing the scaffolds.

With this composite strategy, there are two approaches for making bioceramic-polymer composite scaffolds:

- Incorporating bioceramic particles in the scaffold through a variety of techniques;
- Coating a polymer scaffold with a thin layer of apatite through biomimetic processes.

Both strategies have been employed in our efforts to develop usable scaffolds for bone tissue engineering.

HYBRID BIOMATERIALS FOR BONE TISSUE ENGINEERING

Tissue engineering has evolved from the use of biomaterials for bone substitution that fulfill the clinical demands of biocompatibility, biodegradability, non-immunogeneity, structural strength and porosity. Porous scaffolds have been developed in many forms and materials, but few reached the need of adequate physical, biological and mechanical properties. In the present paper we report the preparation of hybrid porous polyvinyl alcohol (PVA)/bioactive glass through the sol-gel route, using partially and fully hydrolyzed polyvinyl alcohol, and perform structural characterization. Hybrids containing PVA and bioactive glass with composition $58SiO_2$-$33CaO$-$9P_2O_5$ were synthesized by foaming a mixture of polymer solution and bioactive glass sol-gel precursor solution.

Sol-gel solution was prepared from mixing tetraethoxysilane (TEOS), triethylphosphate (TEP), and calcium chloride as chemical precursors. The hybrid composites obtained after aging and drying at low temperature were chemically and morphologically characterized through infrared spectroscopy and scanning electron microscopy. The degree of hydrolysis of PVA, concentration of PVA solution and different PVA-bioglass composition ratios affect the synthesis procedure. Synthesis parameters must be very well combined in order to allow foaming and gelation. The hybrid scaffolds obtained exhibited macroporous structure with pore size varying from 50 to 600 μm.

Tissue engineering represents a new field that aims to grow complex, three-dimensional tissues or organs to replace damaged tissues. For this purpose, investigation of novel biomaterials for bone engineering represents an essential area for planning tissue engineering approaches. Next generation biomaterials should combine bioactive and bioresorbable properties to activate in vivo mechanisms of tissue regeneration, stimulating the body to heal itself and to facilitate replacement of the scaffold by the regenerating tissue. A variety of biomaterials, including synthetic polymers, ceramics and natural polymers are being used to fabricate synthetic scaffold that acts as guide and stimulus for the three-dimensional tissue growth.

Controlled porous architecture, high porosity, adequate pore size and interconnectivity, of these scaffolds is necessary to facilitate cell seeding and diffusion throughout the whole structure of both cells and nutrients. The minimum pore diameter required for bone ingrowth and angiogenesis into a scaffold is considered to be 100 μm. Previous in vitro work has suggested that the ideal pore diameter for bone ingrowth is between 300-400 μm. The material must exhibit good biocompatibility meaning that the material must not demonstrate severe immunogenicity or cytotoxicity. The ability to control scaffold degradation and mechanical integrity is also a requirement to be fulfilled in the development of optimal scaffolds. Ideally, scaffolds should closely match the properties of the tissue it is to replace.

Either biodegradable polymer or ceramic scaffolds, commonly being considered for bone tissue engineering, lack adequate mechanical properties. Synthetic bioresorbable polymers are easily fabricated into complex structures, yet they are too weak to meet the demands of surgery and the in vivo physiological environment. Conversely, ceramic scaffolds have low toughness and strength. Efforts have been made in developing composite of polymers and ceramics with the aim to increase the mechanical stability and improve tissue interaction.

The material composition might give favorable mechanical properties, strength via ceramic phase, toughness and plasticity via polymer phase. Thus, hybrid materials derived from the integration of biodegradable polymers with bioactive inorganic material might be promising on scaffolds development. The bioactive ceramic phase may be a calcium phosphate ceramic, a glass-ceramic or a bioactive glass. Bioactive glasses are able to form a bond to living bone, which has been attributed to the formation of hydroxyl carbonate apatite layer under physiological conditions. However bioactive glass is a stiff and brittle material, difficult to form into complex shapes and prone to catastrophic fracture under loads. The sol-gel method has been employed for the synthesis of hybrid organic-inorganic composite materials. Sol-gel derived bioactive glass foams were also successfully obtained. In vitro cell studies in the presence of these foams have shown an increase in osteoblast proliferation and collagen production as well as the stimulation of the formation and mineralisation of bone nodules reinforcing their potential. However, their compressive strength is in the range of 0.3 to 2.5 MPa, depending on sintering temperature and final pore structure, a value lower than even trabecular bone.

Nevertheless, it is in the range obtained for several scaffolds for tissue engineering reported in the literature. However the toughness and tensile strength of the foams are lower than those of bone. It is expected that the preparation of polymer bioactive glass hybrid foams may lead to scaffolds with better mechanical behaviour compared to pure bioactive glass foams. It should be emphasized however that these materials are expected to be used in sites free of dynamic load, or with controlled load during tissue regeneration.

Polyvinyl alcohol, PVA, is a water soluble synthetic resin which is obtained through polymerization of vinyl acetate monomer. By hydrolysis, the acetate groups are converted in hydroxyl groups. The degree of hydrolysis in a polyvinyl alcohol reagent is controlled by this process. The polar nature of poly (vinyl alcohol) facilitates the formation of hydrogen bonds and eventual condensation with silanol groups (from developing polysilicate network) formed by hydrolysis of the silicon alkoxides.

Moreover PVA has been proposed for controlled release systems and is employed in a variety of biomedical applications, generally being considered to be biocompatible. Although not a biodegradable polymer itself, when used with molecular weight less than around 10.000 g/mol and associated with a biodegradable sol-gel derived bioactive glass, PVA molecules are expected to be eliminated by the body.

Synthetic bioactive and bioresorbable composite materials are becoming increasingly important as scaffolds for tissue engineering. Bioactive glass reacts with physiological fluids to form tenacious bonds to hard (and in some cases soft) tissue. Composites of tailored physical, biological and mechanical properties as well as predictable degradation behaviour can be produced combining bioresorbable polymers and bioactive inorganic phases. Both PVA, as well as bioactive glasses, are being currently used as biomaterials, candidates for tissue engineering applications. In previous work of our group the synthesis of PVA/silica hybrids obtained via sol-gel method was reported. The hybrids were prepared with low polymer contents (20-40 wt. (per cent)).

In this work, we present results concerning the sol-gel synthesis of polyvinyl alcohol-bioactive glass hybrid foams, with polymer contents from 60 to 80 wt. (per cent), and different types of PVA. The purpose of this study was to investigate the effects of some important synthesis parameters such as degree of hydrolysis of PVA, concentration of PVA solution and higher PVA content on the final properties of composites. As far as we know, this is the first research in this technology field to address the influence of PVA chemical properties such as degree of hydrolysis on the fabrication of bioactive glass-polymer (PVA) hybrid tri-dimensional scaffolds.

Experimental

Hybrids Preparation

Sol-gel derived bioactive glass/polyvinyl alcohol hybrid foams were prepared using a procedure similar to the one used by Pereira et al.

Preparation of PVA Solution

The Polyvinyl alcohol (PVA) selected for use were: from Celanese Chemicals, Celvol 103, degree of hydrolysis 98.0-98.8 per cent, molecular weight range: 13,000-23,000 g/mol and from Aldrich-Sigma, degree of hydrolysis 80 per cent, molecular weight: 9,000-10,000 g/mol. PVA aqueous solutions were prepared at 23, 28 and 35 wt. (per cent) by dissolving the PVA powder in a water bath at 80°C, under constant stirring, for 2 hours. The pH of the solution was adjusted to 2.0 by hydrochloric acid solution 2 N.

Preparation of the Starting Bioglass Solution

The starting sol solution with composition 58 wt. (per cent) SiO_2- 33 wt. (per cent) CaO- 9 wt. (per cent) P2O5 was synthesized from mixing tetraethoxysilane (TEOS), D.I. water, triethylphosphate (TEP), and calcium chloride in presence of hydrochloric acid solution 2 N. The H2O/TEOS molar ratio used was 12.

TISSUE ENGINEERING IN COMBINATION OF CELLS

Tissue engineering is the use of a combination of cells, engineering and materials methods, and suitable biochemical and physio-chemical factors to improve or replace biological functions. While most definitions of tissue engineering cover a broad range of applications, in practice the term is closely associated with applications that repair or replace portions of or whole tissues (i.e., bone, cartilage, blood vessels, bladder, etc.). Often, the tissues involved require certain mechanical and structural properties for proper functioning.

The term has also been applied to efforts to perform specific biochemical functions using cells within an artificially-created support system (e.g. an artificial pancreas, or a bioartificial liver). The term regenerative medicine is often used synonymously with tissue engineering, although those involved in regenerative medicine place more emphasis on the use of stem cells to produce tissues.

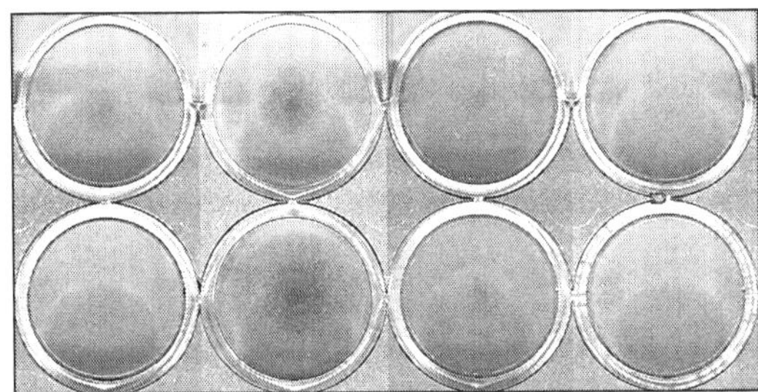

A commonly applied definition of tissue engineering, as stated by Langer and Vacanti, is "an interdisciplinary field that applies the principles of engineering and life sciences toward the development of biological substitutes that restore, maintain, or improve tissue function or a whole organ". Tissue engineering has also been defined as "understanding the principles of tissue growth, and applying this to produce functional replacement tissue for clinical use." A further description goes on to say that an "underlying supposition of tissue engineering is that the employment of natural biology of the system will allow for greater success in developing therapeutic strategies aimed at the replacement, repair, maintenance, and/or enhancement of tissue function." Powerful developments in the multidisciplinary field of tissue engineering have yielded a novel set of tissue replacement parts and implementation strategies. Scientific advances in biomaterials, stem cells, growth and differentiation factors, and biomimetic environments have created unique opportunities to fabricate tissues in the laboratory from combinations of engineered extracellular matrices ("scaffolds"), cells, and biologically active molecules.

Among the major challenges now facing tissue engineering is the need for more complex functionality, as well as both functional and biomechanical stability in laboratory-grown tissues destined for transplantation. The continued success of tissue engineering, and the eventual development of true human replacement parts, will grow from the convergence of engineering and basic research advances in tissue, matrix, growth factor, stem cell, and developmental biology, as well as materials science and bioinformatics. The NSF published a report entitled "The Emergence of Tissue Engineering as a Research Field", which gives a thorough description of the history of this field.

Examples

- Bioartificial liver device — several research efforts have produced hepatic assist devices utilizing living hepatocytes.
- Artificial pancreas — research involves using islet cells to produce and regulate insulin, particularly in cases of diabetes.
- Artificial bladders — Anthony Atala has successfully implanted artificially grown bladders into seven out of approximately 20 human test subjects as part of a long-term experiment.

- Cartilage — lab-grown tissue was successfully used to repair knee cartilage.
- Doris Taylor's heart in a jar
- Tissue-engineered airway
- Artificial skin constructed from human skin cells embedded in collagen
- Artificial bone marrow

CELLS AS BUILDING BLOCKS

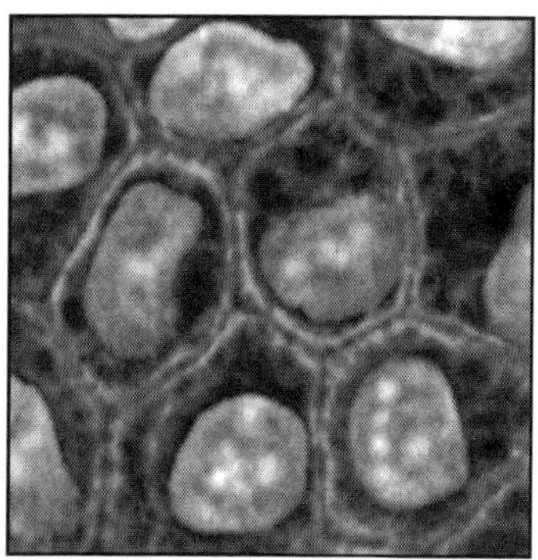

Fig. Stained Cells in Culture

Tissue engineering utilizes living cells as engineering materials. Examples include using living fibroblasts in skin replacement or repair, cartilage repaired with living chondrocytes, or other types of cells used in other ways. Cells became available as engineering materials when scientists at Geron Corp. discovered how to extend telomeres in 1998, producing immortalized cell lines. Before this, laboratory cultures of healthy, noncancerous mammalian cells would only divide a fixed number of times, up to the Hayflick limit.

Extraction

From fluid tissues such as blood, cells are extracted by bulk methods, usually centrifugation or apheresis. From solid tissues, extraction is more difficult. Usually the tissue is minced, and then digested with the enzymes trypsin or collagenase to remove the extracellular matrix that holds the cells. After that, the cells are free floating, and extracted using centrifugation or apheresis. Digestion with trypsin is very dependent on temperature. Higher temperatures digest the matrix faster, but create more damage. Collagenase is less temperature dependent, and damages fewer cells, but takes longer and is a more expensive reagent.

Types of Cells

Cells are often categorized by their source:

- Autologous cells are obtained from the same individual to which they will be reimplanted. Autologous cells have the fewest problems with rejection and pathogen transmission, however in some cases might not be available. For example in genetic disease suitable autologous cells are not available. Also very ill or elderly persons, as well as patients suffering from severe burns, may not have sufficient quantities of autologous cells to establish useful cell lines.

Moreover since this category of cells needs to be harvested from the patient, there are also some concerns related to the necessity of performing such surgical operations that might lead to donor site infection or chronic pain. Autologous cells also must be cultured from samples before they can be used: this takes time, so autologous solutions may not be very quick. Recently there has been a trend towards the use of mesenchymal stem cells from bone marrow and fat. These cells can differentiate into a variety of tissue types, including bone, cartilage, fat, and nerve. A large number of cells can be easily and quickly isolated from fat, thus opening the potential for large numbers of cells to be quickly and easily obtained. Several companies have been founded to capitalize on this technology, the most successful at this time being Cytori Therapeutics.

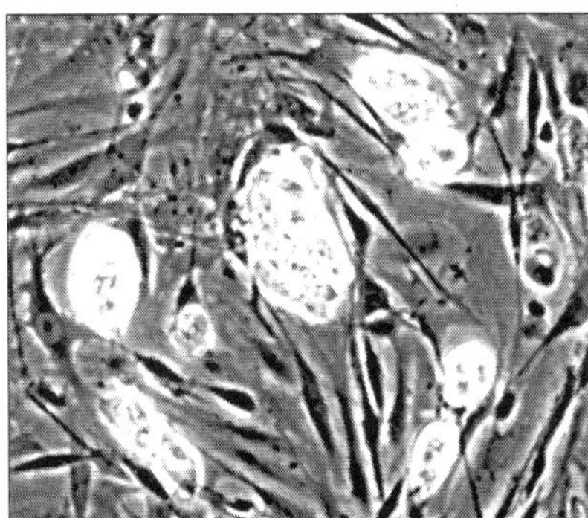

Fig. Mouse Embryonic Stem Cells. More Lab Photos

- Allogenic cells come from the body of a donor of the same species. While there are some ethical constraints to the use of human cells for in vitro studies, the employment of dermal fibroblasts from human foreskin has been demonstrated to be immunologically safe and thus a viable choice for tissue engineering of skin.
- Xenogenic cells are these isolated from individuals of another species. In particular animal cells have been used quite extensively in experiments aimed at the construction of cardiovascular implants.

- Syngenic or isogenic cells are isolated from genetically identical organisms, such as twins, clones, or highly inbred research animal models.
- Primary cells are from an organism.
- Secondary cells are from a cell bank.
- Stem cells are undifferentiated cells with the ability to divide in culture and give rise to different forms of specialized cells. According to their source stem cells are divided into "adult" and "embryonic" stem cells, the first class being multipotent and the latter mostly pluripotent; some cells are totipotent, in the earliest stages of the embryo. While there is still a large ethical debate related with the use of embryonic stem cells, it is thought that stem cells may be useful for the repair of diseased or damaged tissues, or may be used to grow new organs.

Scaffolds

Cells are often implanted or 'seeded' into an artificial structure capable of supporting three-dimensional tissue formation. These structures, typically called scaffolds, are often critical, both ex vivo as well as in vivo, to recapitulating the in vivo milieu and allowing cells to influence their own microenvironments. Scaffolds usually serve at least one of the following purposes:

- Allow cell attachment and migration
- Deliver and retain cells and biochemical factors
- Enable diffusion of vital cell nutrients and expressed products
- Exert certain mechanical and biological influences to modify the behaviour of the cell phase

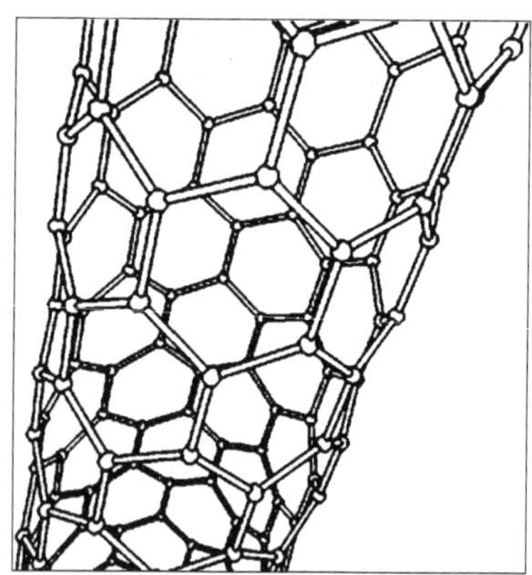

Fig. Carbon Nanotube

Carbon nanotubes are among the numerous candidates for tissue engineering scaffolds since they are biocompatible, resistant to biodegradation and can be functionalized with biomolecules. However, the possibility of toxicity with non-biodegradable nano-materials is not fully understood. To achieve the goal of tissue reconstruction, scaffolds must meet some specific requirements. A high porosity and an adequate pore size are necessary to facilitate cell seeding and diffusion throughout the whole structure of both cells and nutrients. Biodegradability is often an essential factor since scaffolds should preferably be absorbed by the surrounding tissues without the necessity of a surgical removal.

The rate at which degradation occurs has to coincide as much as possible with the rate of tissue formation: this means that while cells are fabricating their own natural matrix structure around themselves, the scaffold is able to provide structural integrity within the body and eventually it will break down leaving the neotissue, newly formed tissue which will take over the mechanical load. Injectability is also important for clinical uses.

Materials

Many different materials (natural and synthetic, biodegradable and permanent) have been investigated. Most of these materials have been known in the medical field before the advent of tissue engineering as a research topic, being already employed as bioresorbable sutures. Examples of these materials are collagen or some linear aliphatic polyesters. New biomaterials have been engineered to have ideal properties and functional customization: injectability, synthetic manufacture, biocompatibility, non-immunogenicity, transparency, nano-scale fibers, low concentration, resorption rates, etc. PuraMatrix, originating from the MIT labs of Zhang, Rich, Grodzinsky and Langer is one of these new biomimetic scaffold families which has now been commercialized and is impacting clinical tissue engineering.

A commonly used synthetic material is PLA - polylactic acid. This is a polyester which degrades within the human body to form lactic acid, a naturally occurring chemical which is easily removed from the body. Similar materials are polyglycolic acid (PGA) and polycaprolactone (PCL): their degradation mechanism is similar to that of PLA, but they exhibit respectively a faster and a slower rate of degradation compared to PLA.Scaffolds may also be constructed from natural materials: in particular different derivatives of the extracellular matrix have been studied to evaluate their ability to support cell growth.

Proteic materials, such as collagen or fibrin, and polysaccharidic materials, like chitosan or glycosaminoglycans (GAGs), have all proved suitable in terms of cell compatibility, but some issues with potential immunogenicity still remains. Among GAGs hyaluronic acid, possibly in combination with cross linking agents (e.g. glutaraldehyde, water soluble carbodiimide, etc...), is one of the possible choices as scaffold material. Functionalized groups of scaffolds may be useful in the delivery of small molecules (drugs) to specific tissues.

Synthesis

A number of different methods has been described in literature for preparing porous structures to be employed as tissue engineering scaffolds. Each of these techniques presents its own advantages, but none is devoid of drawbacks.

- *Nanofiber Self-Assembly:* Molecular self-assembly is one of the few methods to create biomaterials with properties similar in scale and chemistry to that of the natural in vivo extracellular matrix (ECM). Moreover, these hydrogel scaffolds have shown superior in vivo toxicology and biocompatibility compared with traditional macroscaffolds and animal-derived materials.
- *Textile technologies:* these techniques include all the approaches that have been successfully employed for the preparation of non-woven meshes of different polymers.
  In particular non-woven polyglycolide structures have been tested for tissue engineering applications: such fibrous structures have been found useful to grow different types of cells. The principal drawbacks are related to the difficulties of obtaining high porosity and regular pore size.
- *Solvent Casting & Particulate Leaching (SCPL):* this approach allows the preparation of porous structures with regular porosity, but with a limited thickness. First the polymer is dissolved into a suitable organic solvent (e.g. polylactic acid could be dissolved into dichloromethane), then the solution is cast into a mold filled with porogen particles. Such porogen can be an inorganic salt like sodium chloride, crystals of saccharose, gelatin spheres or paraffin spheres. The size of the porogen particles will affect the size of the scaffold pores, while the polymer to porogen ratio is directly correlated to the amount of porosity of the final structure. After the polymer solution has been cast the solvent is allowed to fully evaporate, then the composite structure in the mold is immersed in a bath of a liquid suitable for dissolving the porogen: water in case of sodium chloride, saccharose and gelatin or an aliphatic solvent like hexane for paraffin. Once the porogen has been fully dissolved a porous structure is obtained.
  Other than the small thickness range that can be obtained, another drawback of SCPL lies in its use of organic solvents which must be fully removed to avoid any possible damage to the cells seeded on the scaffold.
- *Gas Foaming*: to overcome the necessity to use organic solvents and solid porogens a technique using gas as a porogen has been developed. First disc shaped structures made of the desired

polymer are prepared by means of compression molding using a heated mold. The discs are then placed in a chamber where are exposed to high pressure CO_2 for several days. The pressure inside the chamber is gradually restored to atmospheric levels. During this procedure the pores are formed by the carbon dioxide molecules that abandon the polymer, resulting in a sponge like structure. The main problems related to such a technique are caused by the excessive heat used during compression molding (which prohibits the incorporation of any tempe-rature labile material into the polymer matrix) and by the fact that the pores do not form an interconnected structure.

- *Emulsification/Freeze-drying:* this technique does not require the use of a solid porogen like SCPL. First a synthetic polymer is dissolved into a suitable solvent (e.g. polylactic acid in dichloromethane) then water is added to the polymeric solution and the two liquids are mixed in order to obtain an emulsion. Before the two phases can separate, the emulsion is cast into a mold and quickly frozen by means of immersion into liquid nitrogen.

  The frozen emulsion is subsequently freeze-dried to remove the dispersed water and the solvent, thus leaving a solidified, porous polymeric structure. While emulsification and freeze-drying allows a faster preparation if compared to SCPL, since it does not require a time consuming leaching step, it still requires the use of solvents, moreover pore size is relatively small and porosity is often irregular. Freeze-drying by itself is also a commonly employed technique for the fabrication of scaffolds. In particular it is used to prepare collagen sponges: collagen is dissolved into acidic solutions of acetic acid or hydrochloric acid that are cast into a mold, frozen with liquid nitrogen then lyophilized.
- *Thermally Induced Phase Separation (TIPS):* similar to the previous technique, this phase separation procedure requires the use of a solvent with a low melting point that is easy to sublime. For example dioxane could be used to dissolve polylactic acid, then phase separation is induced through the addition of a small quantity of water: a polymer-rich and a polymer-poor phase are formed. Following cooling below the solvent melting point and some days of vacuum-drying to sublime the solvent a porous scaffold is obtained. Liquid-liquid phase separation presents the same drawbacks of emulsification/freeze-drying.
- *CAD/CAM Technologies:* since most of the approaches are limited when it comes to the control of porosity and pore size, computer

assisted design and manufacturing techniques have been introduced to tissue engineering. First a three-dimensional structure is designed using CAD software, then the scaffold is realized by using ink-jet printing of polymer powders or through Fused Deposition Modeling of a polymer melt.

Assembly Methods

One of the continuing, persistent problems with tissue engineering is mass transport limitations. Engineered tissues generally lack an initial blood supply, thus making it difficult for any implanted cells to obtain sufficient oxygen and nutrients to survive, and/or function properly. Self-assembly may play an important role here, both from the perspective of encapsulating cells and proteins, as well as creating scaffolds on the right physical scale for engineered tissue constructs and cellular ingrowth. It might be possible to print organs, or possibly entire organisms. A recent innovative method of construction uses an ink-jet mechanism to print precise layers of cells in a matrix of thermoreversable gel. Endothelial cells, the cells that line blood vessels, have been printed in a set of stacked rings. When incubated, these fused into a tube.

Tissue Culture

In many cases, creation of functional tissues and biological structures in vitro requires extensive culturing to promote survival, growth and inducement of functionality. In general, the basic requirements of cells must be maintained in culture, which include oxygen, pH, humidity, temperature, nutrients and osmotic pressure maintenance. Tissue engineered cultures also present additional problems in maintaining culture conditions. In standard cell culture, diffusion is often the sole means of nutrient and metabolite transport. However, as a culture becomes larger and more complex, such as the case with engineered organs and whole tissues, other mechanisms must be employed to maintain the culture. Another issue with tissue culture is introducing the proper factors or stimuli required to induce functionality. In many cases, simple maintenance culture is not sufficient.

Growth factors, hormones, specific metabolites or nutrients, chemical and physical stimuli are sometimes required. For example, certain cells respond to changes in oxygen tension as part of their normal development, such as chondrocytes, which must adapt to low oxygen conditions or hypoxia during skeletal development. Others, such as endothelial cells, respond to shear stress from fluid flow, which is encountered in blood vessels.

PRECURSORS OF TISSUE ENGINEERING

Tissue engineering represents the confluence of a complex array of pre-existing lines of work from three quite different domains: the worlds of clinical medicine, engineering, and science. This section offers a brief overview of the range and character of these different contributing elements, to provide a context for the discussion in later sections of intellectual and institutional developments in TE.

ROOTS IN CLINICAL MEDICINE

Perhaps the most obvious precursors to TE lie in the clinical domain. These are best understood as specific examples of general problem-solving strategies employed by physicians. Consider, for example, the basic dilemma faced by a surgeon. While the removal of organs or body structures that are damaged beyond repair by disease or trauma can be life-saving, the patient must cope with the functional effects of tissue loss and, in some cases, the psychological impacts of disfigurement. And for those vital organs whose complete removal is incompatible with life, the surgeon's hard-earned skill is, by itself, of no avail: the procedure cannot be done unless there is some way of replacing or reconstituting essential functions.

The resourceful application of virtuosic craft through the tools and materials at hand to meet the distinctive needs of each patient is central to the ethos of surgery. Thus, surgeons have sought to reconstruct anatomic structures using the patient's own tissues as raw material; they have pressed artificial materials into service as prostheses; and, most spectacularly, they have brought patients back from the brink of death by transplanting an ever-wider range of vital organs – primarily living organs, but in a few cases, with only very limited success to date, prototype artificial organs as well.

However, with experience, surgeons have come to understand in detail not only the benefits of such measures, but their limitations as well. Anatomic reconstruction using the patient's own tissues can cause substantial morbidity at the donor site; the improvised structures are usually functionally inferior to the natural organs they replace, and less durable as well. Poor compatibility between artificial materials and mechanical systems and the internal environment and physiologic requirements of the human body can lead to dysfunctional interactions and new failure modes. Transplantation of living organs brings with it profound immunologic complications, and the number of patients who can be treated in this way will always be severely constrained by the limited supply of organs suitable for use.

For a surgeon, then, the development of engineered tissues is a logical next step in the ongoing effort to improve the match between the surgeon's various reparative and reconstructive contrivances and the requirements of human anatomy and physiology. Physicians in a range of internal medicine specialties as well have found themselves impelled to explore clinical solutions that incorporate living cells. Generally, internists seek to identify therapies that can repair or reconstitute physiologic functions with sufficient effectiveness to enable patients to avoid surgery. Often these therapies are pharmacologic – physicians may use small molecules, or, increasingly, complex, genetically-engineered biological macromolecules, to replace critical molecular species that are in short supply within the body, to counter the effects of molecules that are in harmful oversupply, or to intervene in more subtle ways in regulatory pathways that control critical functions. Other therapies rely on physico-chemical effects – sometimes implemented through external artificial devices – to replace critical filtering functions that maintain electrolyte balance and remove metabolic wastes from the body.

Yet the metabolic functions of human organs are so complex and interrelated that simple pharmacologic or physico-chemical approaches, while they can be life-saving, nevertheless often constitute highly imperfect solutions whose limitations become increasingly apparent in chronic use. In such situations, the notion of leveraging the kinds of complex, integrated physiologic functions that are accessible only at the level of intact cells or tissues becomes compelling.

Examples

Landmark conceptual developments prerequisite for a concept of tissue engineering have emerged over a period of decades within the problem-solving traditions of clinical medicine. In some clinical domains, physicians and non-clinician researchers reached the stage of incorporating living cells into prototype tissue-engineered clinical solutions years before the emergence of a generalized concept of tissue engineering.

Vascular grafts

The earliest explorations by surgeons of the possibility of transplanting blood vessels took place more than a century ago; the renowned surgical researcher Alexis Carrel was awarded the 1912 Nobel prize in physiology or medicine for his demonstration of successful techniques for the anastomosis of blood vessels and the extension of these techniques from the transplantation of vessels to the transplantation of entire solid organs. Over the succeeding decades, experimental use of rigid glass and metal tubes as vascular grafts yielded disappointing results. In the early 1950s, Voorhees demonstrated the first use of tubes of synthetic fabric as arterial prostheses.

With the expanded use of a range of synthetic grafts in clinical practice and ongoing research into the characteristics of a range of alternative materials, surgeons and biomaterials researchers gained a deeper appreciation of thrombogenesis and other problems arising from the interaction between synthetic materials and the blood and perigraft tissues with which they came in contact. The concept of a resorbable vascular graft was introduced in the 1960s, and the first fully-resorbable graft was reported in 1979. Improvement in the healing process of Dacron vascular grafts via pre-seeding with endothelial cells was reported in 1978. Finally, the first attempt to create entirely biologic vascular structures in vitro, using collagen and cultured vascular cells, was reported in 1982.

Skin grafts

For centuries, physicians have attempted to cover severe wounds with grafts from a variety of sources, including cadavers and living humans. By the early decades of the 20th century, researchers were investigating the immunologic basis for rejection of skin allografts, though there was no apparent progress toward any practical solution.

The marked increase during World War II in the number of burn victims for whom a skin allograft was not feasible provided a renewed impetus for research on skin replacement. During this era, the distinguished immunologist Peter Medawar made important contributions both to further progress in the understanding of the immunology of graft rejection, and to the in vitro culture of epithelial cells drawn from a patient. By 1953, Billingham and Reynolds demonstrated in animal models that the products of a brief culture of epidermal cells could be applied to a graft bed to reconstitute an epidermis.

Despite these early successes, more efficient means of cultivation were needed to provide enough cells to sustain transplantation. Overgrowth of certain cell types, such as fibroblasts, suggested that existing culture techniques would not be effective in producing large quantities of cells. Research in the 1960s and 70s identified growth factors that could be added to culture medium to induce greater proliferation of epidermal cells. Starting in the mid-1970s, three research groups working independently at MIT reported a series of milestones in the development of skin replacements.

In 1975, Green and Rheinwald described the co-culture method, a technique for serial cultivation of human epidermal keratinocytes from small biopsy samples.Using the technique, one sample of cells was sufficient to generate enough thick, multilayered skin to resurface the entire body of a burn victim. Some differentiation among epidermal cells to mimic that of the epidermis was also visible. In 1979, Green and colleagues, building on the work of Bellingham and Reynolds, demonstrated that cultured cells can be grown in sheets in a petri dish and transferred intact, rather than as disaggregated cells, to a graft wound bed. That same year, Bell and colleagues described the use of fibroblasts to condense a hydrated collagen lattice to a tissue-like structure potentially suitable for wound healing.

These findings led to the first functional living skin equivalent (LSE) in 1981, consisting of fibroblasts suspended in a collagenglycosaminoglycan matrix. Yannas identified the components of the underlying matrix structure of skin and used this knowledge to develop a dermal regeneration template which, when implanted and seeded with autologous basal cells, stimulated re-growth of functional skin.16 All of these lines of research, continuing into the 1990s, proceeded to the development of commercial products.

Kidney

A number of investigators experimented sporadically with kidney transplantation during the early part of the 20th century, but planned renal transplantation efforts began only in the late 1940s. During the war years in the Netherlands, Kolff developed the first dialysis machine; its design was refined at the Peter Bent Brigham Hospital in Boston and used in patients for the first time in 1948. In turn, the availability of effective short-term dialysis facilitated progress on transplantation, culminating in the successful transplant of a donated kidney from a twin by Murray and colleagues at Brigham and Women'a Hospital in 1954. Subsequent advances in immunosuppression for transplantation and further development of Kolff's dialysis machine made both techniques practical for widespread, routine use, transforming the management of end-stage renal disease.

However, the limited supply of organs suitable for transplantation, combined with recognition of the therapeutic limitations of the dialysis machine, motivated the formulation of the concept of the bioartificial kidney, which would mimic more faithfully the physiologic functions of the kidney and thus avoid the debilitating side effects of chronic dialysis. During the late 1960s and early 1970s, Wolf experimented with combinations of kidney cells with hollow synthetic fibers as conduits for nutrients and waste.

Subsequent work on growing liver cells on the outside of hollow fibers by Wolf and independently by Knazek led to the demonstration of hollow-fibre bioreactors. In the mid-1980s Galletti and colleagues furthered development of the bioartificial kidney concept through their research on hollow-fibre bioreactors employing renal epithelial cells.

Pancreas/ islet cells

Although the introduction of insulin more than 70 years ago had a miraculous impact on the lives of diabetes patients, with prolonged survival it became apparent that the highly imperfect glycemic control typical of routine insulin therapy was associated with severe long-term complications for a variety of organ systems. The first insulin pumps appeared in the 1960s, but the sensor and control technology required for a complete "closed loop" system to mimic the adaptive character of physiologic glucose control was not available. The first pancreas transplant, in conjunction with a simultaneous kidney transplant, was performed by Lillehei in 1966.

Lacy reported a method for isolation of intact islets in 1967, and isolated islet cells were first transplanted in 1970, though without a solution to the problem of immune rejection. The use of microencapsulated islets as artificial beta cells was proposed by Chang as early as the mid-1960s. During the 1970s,

Chick and colleagues, building on the work of Knazek, developed a "hybrid artificial pancreas" consisting of beta cells cultured on synthetic semipermeable hollow fibers, and demonstrated the ability of this device to restore glucose homeostasis in rats when connected to the circulatory system via shunt. Sun and colleagues reported similar work in the 1970s, followed by studies of implanted microencapsulated islets beginning in 1980. Investigation of different ways of "packaging" islet cells to provide effective and durable glycemic control continued through the 1980s and beyond.

Liver

The first successful liver transplant was carried out by Starzl in 1967, but in the absence of an adequate supply of transplantable organs, researchers and clinicians have continued to pursue alternative approaches to the replacement of hepatic function. Over more than four decades, clinicians have attempted in a variety of ways to provide extracorporeal support to patients suffering from liver failure. Nonbiological approaches that have been explored include hemodialysis, hemoperfusion over charcoal or resins or immobilized enzymes, plasmapheresis, and plasma exchange.

However, these approaches have met with limited success, presumably because the complex synthetic and metabolic functions of the liver are inadequately replaced by these systems. Work on cell-based therapies and bioartificial systems has reflected the same themes observed in the case of pancreatic islet cells. Wolf and Munkelt reported in 1975 on a bioreactor containing a rat hepatoma cell line cultured on the surface of semipermeable hollow fibers within a plastic housing. Sutherland and colleagues reported in 1977 on the use of transplanted hepatocytes in the treatment of drug-induced liver failure in rats, and Sun and colleagues in the mid- 1980s explored approaches to microencapsulation of hepatocytes.

Bone and Cartilage

A variety of materials generally perceived as chemically inert, such as various metals and alloys, have been used for many years to replace damaged bone or to provide support for healing bones. With experience, however, it has become clear that non-biologic materials do not remain biologically inert in the environment of the human body, but rather elicit reactions whose intensity is related to a variety of factors such as implantation site, the type of trauma at the time of surgery, and the precise material in use. Beginning in the 1970's, bioactive materials, such as porous glass and hydroxyapatite ceramic, were examined as alternatives, as they were shown to elicit the formation of normal tissue on their surfaces.

Aside from novel biomaterial development, the growth and regenerative capacities inherent in bone have also been intense topics of scientific and clinical study for decades. In a 1945 publication in Nature, Lacroix hypothesized that osteogenin, a substance in bone, was responsible for its growth. Twenty years later in 1965, Marshall Urist proved that there was, indeed, some substance or combination of substances present in demineralized bone, which when transplanted, could induce growth of new bone. Urist's landmark finding encouraged many to investigate the precise factors which trigger bone induction.

During the 1970s and 80s, research demonstrated that the process is mediated by a category of growth factors, termed bone morphogenetic proteins or BMPs, which act in a multistep cascade highly reminiscent of embryonic bone morphogenesis. Reddi and colleagues developed techniques to isolate these proteins from the extracellular matrix of bone. The findings are promising, too, for induction of cartilage as BMPs initially induce a cascade of chondrogenesis and might just as easily be called cartilage morphogenetic proteins. Work on isolation, purification, and proliferation of BMPs continued throughout the 1980s.

EMERGENCE AND EVOLUTION OF A SHARED CONCEPT

It is unclear who first used the term "tissue engineering" to mean what it does today. Not surprisingly for a coinage that seems so natural in hindsight, a number of the individuals interviewed for this study suggested that the term may have been invented several times independently before it came into sufficiently broad usage that a wide range of researchers can be expected to have encountered it in publications or in discussion. Indeed, the first appearance of the term in print of which the study team is aware – also the earliest revealed through a PubMed search – was an incidental, almost offhand usage in a 1984 publication that described the organization of an endothelium-like membrane on the surface of a long-implanted, synthetic ophthalmic prosthesis.

However, the origin of "tissue engineering" as it is recognized today can be clearly traced to a specific individual. In 1985, Y.C. Fung, a pioneer of the field of biomechanics and of bioengineering more broadly, submitted a proposal to NSF for an Engineering Research Centre to be entitled "Centre for the Engineering of Living Tissues". Fung's concept drew on the traditional definition of "tissue" as a fundamental level of analysis of living organisms, between cells and organs: The study of organs and organ systems has historically been the domain of the physiologist and physician.

There is, therefore, a relative wealth of practical information about organs, codified in terms of medical practice. On the other hand, tissues are composed of cells, having specialized internal organelles and, ultimately, chemical constituents. The composition of the cell and its constituents has been dealt with by cell biologist and biochemist. There are relatively few focused efforts at bridging the gap between these extremes.

A clear understanding of phenomena at the tissue level is prerequisite to the engineering of tissues. Fung's proposal was not accepted. Nevertheless, the concept of an engineering approach to the level of biological organization between cells and organs surfaced again at NSF in the spring of 1987, at a panel meeting convened to review proposals to the Bioengineering and Research to Aid the Handicapped (BRAH) Programme within the Engineering Directorate. Fung was present at this meeting, and is recalled as having volunteered the term "tissue engineering" in the course of a discussion that was seeking to crystallize the concept.

Further discussions between the programme directors for the BRAH Programme, which looked at whole organs, and the Directoratefor Engineering's Biotechnology (BIOTECH) Programme, which focused on the cellular level, led to the convening of a special Panel Meeting on Tissue Engineering at NSF on October 28, 1987. For this meeting, Allen Zelman, a Programme Director for BRAH, prepared a draft definition of tissue engineering:

The term "tissue engineering" indicates a new inter-disciplinary initiative which has the goal of growing tissues or organs directly from a single cell taken from an individual. Interestingly, in his "statement of the problem", Zelman pointed to the avoidance of immune rejection through the growth of tissues or organs from a patient's own cells as the key benefit of tissue engineering. Zelman envisioned the development of a large and vigorous new industry producing internal organs, but tempered this vision with the caveat that production of complex internal organs, such as the kidney, "would be considered far too ambitious as a starting point" and that "tissues, being more simple than organs, should be investigated initially".

Subsequent to this meeting, a Forum on Issues, Expectations, and Prospects for Emerging Technology Initiation was held in Washington, DC, under the sponsorship of the Division of Emerging Engineering Technologies within NSF. This forum recommended that tissue engineering be designated as an emerging engineering technology, and that a workshop be held to identify appropriate areas for research in this technology.

This proposed workshop, organized by Zelman, Frederick Heineken, and Duane Bruley –all of NSF—was held at Granlibakken Resort, Lake Tahoe, California, in February 1988. With the exception of a re-publication of the 1984 ophthalmology paper in another ophthalmology journal in 1985, the next known appearance of the term "tissue engineering" in print was in the proceedings of the Granlibakken workshop. A preface to these proceedings defined the term more broadly than did any of the provisional definitions floated up to that point, to encompass a wider range of potential therapeutic interventions that could be enabled by research carried out under this new perspective: "Tissue Engineering" is the application of principles and methods of engineering and life sciences toward fundamental understanding of structure-function relationships in normal and pathological mammalian tissues and the development of biological substitutes to restore, maintain, or improve tissue function. The basic point of the above definition is that tissue engineering involves the use of living cells plus their extracellular products in development of biological substitutes for replacements as opposed to the use of inert implants. The definition is intended to encompass procedures in which the replacements may consist of cells in suspension, cells implanted on a scaffold such as collagen and cases in which the replacement consists entirely of cells and their extracellular products.

At the 1992 UCLA symposium on tissue engineering, Eugene Bell defined tissue engineering in terms of a more specific list of goals:

- Providing cellular prostheses or replacement parts for the human body;
- Providing formed acellular replacement parts capable of inducing regeneration;
- Providing tissue or organ-like model systems populated with cells for basic research and for many applied uses such as the study of disease states using aberrant cells;
- Providing vehicles for delivering engineered cells to the organism;
- Surfacing non-biological devices.

Langer and Vacanti referenced the definition from the Granlibakken proceedings, presenting it in condensed form: Tissue engineering is an interdisciplinary field that applies the principles of engineering and the life sciences toward the development of biological substitutes that restore, maintain, or improve tissue function. Thus, perhaps the single most cited and influential paper in the field, cites the Granlibakken workshop and builds upon its pioneering definition of the tissue engineering. With this definition as a foundation, they added substance to the notion of common themes underlying a seeming diversity of research by identifying three general strategies for the creation of new tissue – the use of:

- Isolated cells or cell substitutes;
- Tissue-inducing substances; or
- Cells placed on or within matrices.

In the body of the paper, they briefly introduced ongoing efforts across a wide range of organ systems, classified by their embryologic origin – as ectoderm, endoderm, or mesoderm. Finally, they concluded by identifying further common themes, this time in the form of enabling knowledge or technologies of broad significance that should be targets for future research, in the areas of cell biology, cell sourcing and preservation, and materials.

The number of PubMed title/abstract "hits" on the term "tissue engineering" first exceeded 10 in 1994, the year after the Langer/Vacanti review appeared. The number of appearances doubled in 1996, and again in 1998 and 1999, reaching 153 in 1999 and continuing to grow to 214 in 2000. This figure surely understates the extent to which researchers associated the concept "tissue engineering" with their work; once the concept is established, there is no reason for a researcher to call it out explicitly in titles or abstracts unless there is a specific point to be made by doing so. The definitions elaborated during the 1987-93 period provided the basic terms of reference for discussions of tissue engineering through the 1990s.

Tissue engineering researchers seeking to situate their work within a broader context typically cited either the Granlibakken workshop definition or the Langer/Vacanti framing of the field. Even those who did not directly cite these sources offered definitions that reflected some combination of the elements contained in the definitions outlined here, with the exception of Eugene Bell's final point about surfacing non-biological devices, which seems to have been ignored for the most part. These early definitions left at least two important ambiguities. One involved the role of cells in tissue engineering. In many formulations of the concept, the unique aspect of tissue engineering compared to traditional biomedical engineering or to pharmaceutical development was that its products incorporated living cells.

Bell's 1992 definition, allowing for "acellular replacement parts capable of inducing regeneration", reflected an emerging recognition that physiologic reactions to biomaterials were not necessarily just a nuisance to be suppressed, but might under some circumstances offer a means of inducing useful adaptations in the body. However, this conceptual advance also blurred the distinction between this new field and the studies of acellular biomaterials that had been a mainstay of biomedical engineering and materials science. Langer and Vacanti went one step further, defining the use of "tissue-inducing substances" more generally as one of the strategies of tissue engineering. Including growth factors or other "signaling molecules" within the domain of tissue engineering represented progress toward an integrated understanding of the factors that govern tissue development in vivo, but also broke down the boundaries between tissue engineering and modern pharmaceutical research, which draws increasingly on the latest findings of cellular and molecular biologists.

In the words of tissue engineer Jeffrey Hubbell, "Doing tissue engineering with factors to stimulate cells in the body is really just fancy drug delivery. One is delivering a drug – like a protein, a morphogenic factor – that stimulates cellular responses at a site with the goal of ending up with some overall tissue reconstruction or regeneration at that site." A second ambiguity concerned the role of hybrid devices in tissue engineering, and the related question of whether therapeutic products of tissue engineering were necessarily intended to be implanted into the body. Several lines of research typically referred to as tissue engineering pursue the development of external bioreactors that can replace critical metabolic functions. From both a conceptual and a historical perspective, this work arguably represents an incremental advance on the dialysis machine. The longterm vision of researchers working on hybrid devices typically extends to more compact, self-contained versions that can be implanted. However, even when miniaturized, current "bioartificial organs" remain more machines than living organs, closer to today's mechanical artificial heart than to the vision of an adaptive biological implant that is seamlessly incorporated into the body's reparative and homeostatic mechanisms. The linkage of TE research with clinical medicine, although inevitable and desirable in view of the goals of the endeavor, nevertheless has also served as something of an obstacle to the sharpening of a definition of tissue engineering as an academic discipline. For example, orthopedic surgeons have been investigating many different kinds of implant that may promote bone regeneration.

From a clinical perspective, what matters is not so much whether the active agent in an implanted matrix consists of stem cells, bone morphogenetic proteins, or gene therapy vectors, but whether it is therapeutically effective. Similarly, a nephrologist may view the dialysis machine, the external biohybrid "artificial kidney" or functional, histocompatible "microrenal" units created via nuclear transplantation as possibilities along a seamless spectrum of therapeutic options rather than in terms of the radically different underlying technologies they represent. Recent developments in nomenclature reflect this persistent ambiguity in scope and focus.

The National Institutes of Health Bioengineering Consortium (BECON) symposium on tissue engineering, held at NIH in June, was entitled "Reparative Medicine: Growing Tissues and Organs". The proceedings of the symposium offer multiple competing – and to some extent conflicting – definitions of reparative medicine, of tissue engineering, and of the relationship between the two. At one extreme, reparative medicine is defined very broadly as "the replacement, repair, or functional enhancement of tissues and organs", a definition that is not much narrower than all of clinical medicine, although most of the examples cited have a surgical flavour; tissue engineering is viewed as one strategy among many for reparative medicine. At the other, reparative medicine is defined as a synonym for tissue engineering: Reparative medicine, sometimes referred to as regenerative medicine or tissue engineering, is the regeneration and remodeling of tissue in vivo for the purpose of repairing, replacing, maintaining or enhancing organ function, and the engineering and growing of functional tissue substitutes in vitro for implantation in vivo as a biological substitute for damaged or diseased tissues and organs.

This definition of tissue engineering is noteworthy for excluding extracorporeal bioartifical organs, and indeed, such devices gain only a passing mention in the proceedings, which otherwise cover a very broad scope. The term "regenerative medicine", offered as a further synonym, appears to have been coined by William Haseltine, to capture for promotional purposes his view of the future of medicine. Contrary to the usage at the BECON symposium, Haseltine's conception positions TE as a subset – a "thread" or "phase" – of regenerative medicine, not as a synonym for it, emphasizing the in vitro construction of human organs for implantation, using specialized biocompatible materials, signaling molecules, and adult human cells. In practice, however, there is little to distinguish Haseltine's "regenerative medicine" from other conceptions of tissue engineering.

Despite these alternative forms of the term, however, the true sentiment of the field appears to have been captured early on at the NSF meetings of 1987 and 1988. It is this concept of the field, which has been carried on by its leading proponents and remains highly referenced today.

PARTICULATE BIOCERAMICS FOR COMPOSITE SCAFFOLDS

Particulate materials of hydroxyapatite (HA), tricalcium phosphate (TCP) and a few bioactive glasses can be used in composite scaffolds for achieving osteoconductivity. In one of our early investigations, plasma-sprayed HA (psHA) particles were produced and used, as particulate psHA was found to be more bioactive than highly crystalline HA particles due to the amorphous component in psHA particles.

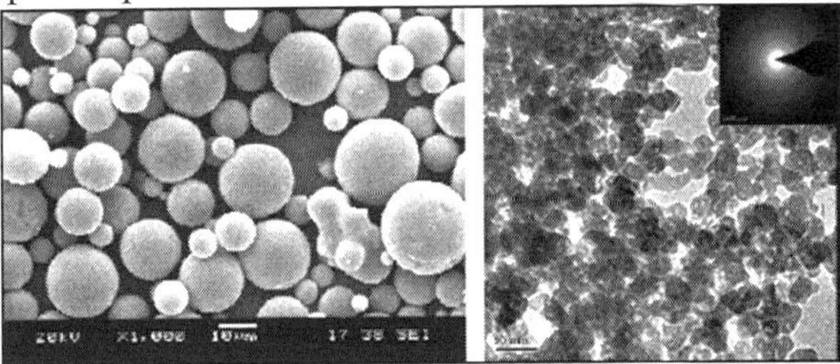

(a) ps HA Microspheres (b) CHA Nanospheres

Fig. Particulate HA for Composite Scaffolds

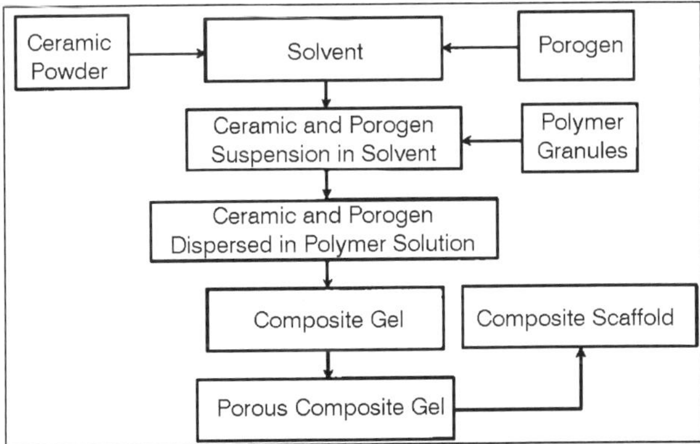

Fig. Manufacture of Bioactve Composite Scaffolds

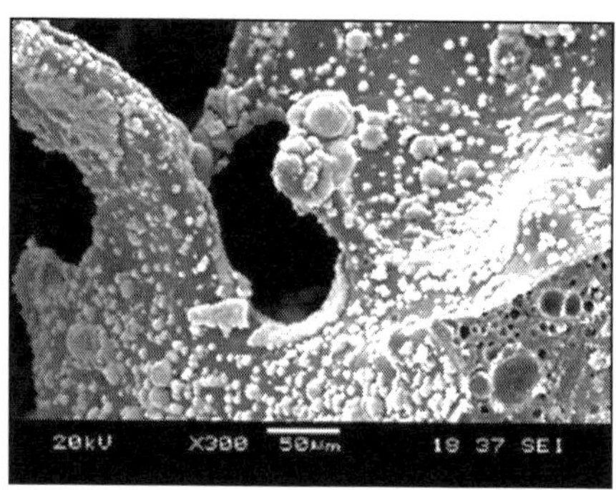

Fig. Formation of bone-like apatite on a ps HA/PLLA Composite Scaffold Through Immersion in Simulated body Fluid

These spherical psHA particles were in the micrometer size range. In our recent investigations, carbonated HA (CHA) particles were synthesized through a nanoemulsion process. CHA is also more soluble and bioactive than highly crystalline HA due to the CO_3 2- substitution in HA lattice. Furthermore, the spherical CHA particles were in the nanometer size range and are thus more suitable for incorporation into polymers to form composite scaffolds.

INCORPORATING BIOCERAMIC PARTICLES IN THE SCAFFOLDS

In our early investigations, psHA particles were incorporated in a poly(L-lactic acid) (PLLA) to form osteoconductive composite scaffolds. In this process, freeze-drying was used at the final stage to produce the porous structure. Just as in the production of polymer scaffolds, for composite scaffolds, factors such as polymer solution concentration, porogen type and size, freeze-drying parameters, etc. play very important roles in forming the scaffolds of desired porous structures (pore geometry, pore size and size distribution, pore interconnectivity, thickness of pore walls, etc.) and hence the mechanical performance. It was shown through in vitro experiments that the composite scaffolds possessed bioactivity due to the incorporation of psHA particles.

Fig. CHA/PHBV Composite Scaffold Produced through an Emulsion Freeze-Drying Process

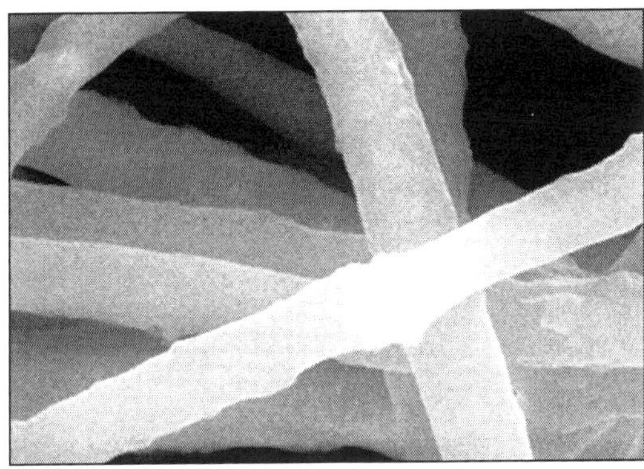

Fig. CHA/PHBV Composite Nanofibers Produced through Electrospinning

In a recent study, it was shown that using an emulsion freeze-drying process, composite scaffolds based on polyhydroxybutyrate (PHB) and polyhydroxybutyrate-co-valerate (PHBV) natural polymers could be produced. In these scaffolds, the CHA nanospheres were incorporated. Through this process, highly porous structure with a controlled pore morphologies could be achieved. In recent years, biodegradable fibers which have diameters in the nanometer range have attracted great attention, as it is considered that the proper in vivo phenotype cannot be consistently achieved if cells are presented with fibers with diameters equal to or greater than the cell size. Among technologies for making polymer fibers of nanometer dimensions, electrospinning is the most appealing and the process is relatively simple. A study was recently conducted on electrospinning of PHB and PHBV polymers and their composites. It was found that electrospinning parameters (polymer solution concentration, solution feeding rate, injection needle diameter, electrical voltage, and working distance) all had significant effects on the resultant fibers.With careful experimentation, CHA/PHBV composite nanofibers containing CHA nanospheres could be produced. The incorporation of CHA nanospheres was confirmed through EDX analyses of the fibers. Rapid prototyping (RP) technologies have become increasingly popular for producing tissue engineering scaffolds, as there are a few distinctive advantages of using RP technologies. One of our efforts is to use the selective laser sintering (SLS) technology to manufacture CHA/PLA composite scaffolds. In the SLS route, there are two major issues:

- Obtaining powders of appropriate characteristics for sintering,
- Minimizing the use of sintering powder.

The latter issue could be solved by building and installing a miniature sintering platform in a commercial SLS machine. The former issue required the production of composite powders of suitable characteristics. Through a microemulsion process, both PLA and CHA/PLA composite powders were made.

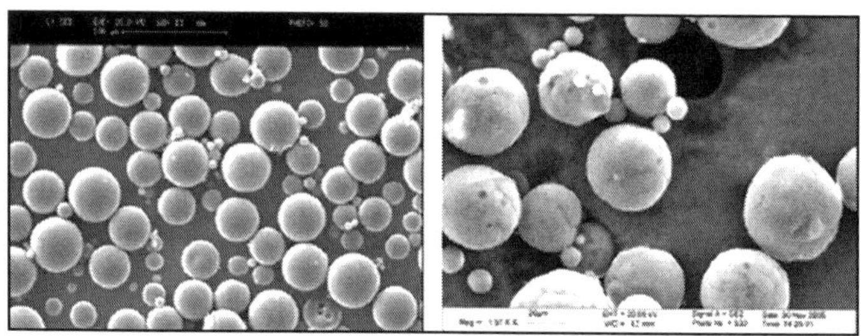

Fig. Powders for Producing Composite Scaffolds through Selective Laser Sintering

BONE TISSUE ENGINEERING USING FOETAL CELL THERAPY

Different cell sources for bone tissue engineering are reviewed. In particular, adult cell source strategies have been based on the implantation of unfractionated fresh bone marrow; purified, culture expanded mesenchymal stem cells, differentiated osteoblasts, or cells that have been modified genetically to express rhBMP. Several limiting factors are mentioned for these strategies such as low number of available cells or possible immunological reaction of the host. Foetal bone cells are presented as an alternative solution and a review of actual treatments using these cells is presented. Finally, foetal cells used specifically for bone tissue engineering are characterised and potentially interesting therapeutic options are proposed.

Bone regeneration is based on the hypothesis that healthy progenitor cells, either recruited or delivered to an injured site, can ultimately regenerate lost or damaged tissue. Three-dimensional bone grafts may enhance bone regeneration by creating and maintaining a space that facilitates progenitor cell migration, proliferation and differentiation. Several techniques have been developed to provide the surgeon with needed material for bone grafts.

Pieces of bone are collected either from the patients undergoing surgery (autograft, which is considered as the gold standard and remains the most used therapy for bone repair) or obtained from a tissue bank (allograft). However, these techniques have some drawbacks such as traumatic procedures, morbidity or increased operating times for autografts and limited supply, risks of contamination or high costs for allografts.

These limitations have led to new research aiming to provide a bone graft engineered in the laboratory and available when needed by the surgeon. The ultimate goal of this new treatment strategy is the regeneration rather than just the repair of skeletal tissue. From a broad point of view, this approach can be defined as "bone tissue engineering". As for any application in tissue engineering, the cell origin and type are essential aspects in bone tissue engineering.

Technically, the cells used should be relatively simple with respect to their collection, culture, expansion and storage. Therapeutically, the cells used should have high bone formation potential, no immunological-induced reactions and no pro-inflammatory properties.

Adult Cell Sources for Bone Tissue Engineering Applications

Cell sources for bone tissue engineering applications can be categorised with respect to their state of differentiation. With this idea, four different cell-based tissue engineering approaches have been described for the regeneration of bone. These strategies are based on the implantation of:

- Unfractionated fresh bone marrow,
- Purified, culture expanded mesenchymal stem cells.
- Differentiated osteoblasts,
- Cells that have been modified genetically to express rhBMP.

Generally, the less differentiated cells will be more easily expanded in vitro due to their high proliferation rate, while the differentiated cells will be more effective in vivo due to their higher production of mineralised extracellular matrix.

For each type of cells used, advantages and disadvantages can be found. The cells from unfractionated fresh bone marrow are relatively easy to collect, but it will not be possible to use these cells in allotransplantation as bone marrow contains T lymphocytes that encounter and respond to host antigens in virtually all tissues in the body, leading to multi-system graft-versus-host syndrome. Mesenchymal stem cells, isolated from bone marrow aspirate, adult peripheral blood, neonatal cord blood or liver for example, could present advantages from an immunological point of view.

However, as one of every 100,000 nucleated cells derived from bone marrow is a stem cell, a procedure of isolation is required in order to decrease the volume of material injected. Compared to unfractionated bone marrow, mesenchymal stem cells have been shown to generate greater bone formation in preclinical studies. However, gradual loss of both their proliferative and differentiation potential has been observed during in vitro expansion.

To overcome this difficulty, telomerase therapy (hTERT) has been recently used and effectively extended stem cell life-span while maintaining or even enhancing their osteogenic potential. Alternative sources of autologous precursor cells were investigated in fat by liposuction or in skeletal muscle by biopsy. It has to be mentioned that pluripotential cells from mesodermal tissues other than bone marrow did not show bone healing in animal models.

The use of predifferentiated osteoblasts has been shown to enhance the rate and extent of bone regeneration. Their expansion in vitro may be however difficult and is highly depending on the donor. Finally, cells genetically modified to express bone formation cytokines could be used to take advantage of genetic therapy, combining gene therapy and tissue engineering methodologies to enhance tissue regeneration. The transfection of the cells can be done either in two steps with an in vitro transfection followed by injection of the modified cells (procedure called ex vivo gene therapy) or in one step by transfecting directly the cells in the body (procedure called in vivo gene therapy). With ex vivo gene therapy approach, cells overexpressing BMP have been developed and used in animal studies. Concerns on possible host immune reaction need to be clarified. A list of requirements for ideal cell sources for bone tissue engineering applications is summarised in table. No adult cells seem to fulfil all the requirements leaving room to propose an alternative approach based on foetal cells.

Foetal Cell Sources for Tissue Engineering Applications

Foetal associated tissues such as placenta, amniotic liquid or umbilical cord are described to be potential sources of cells for tissue engineering. In contrast to embryonic tissue derived up to the end of the 8th week, foetal tissue begins at the 9th week and is considered as an organ donation. The idea of using foetal tissues or cells for cell therapy is mainly coming from the field of neurology. Indeed, several clinical studies have already been performed to treat neurodegenerative disorders. Neuronal affections such as Huntington's or Parkinson's disease have been treated by transplantation of fresh foetal neuroblasts. Unfortunately, these cells are difficult to expand in culture and have to be transplanted freshly therefore needing large quantities of fresh tissue.

Still within the neurology but following accidents, cell therapy for stroke has also tried to benefit from foetal cells, with a clinical trial treating patients with foetal cells obtained from the porcine primordial striatum. In this situation, safety of the treatment was verified, but none of the patients showed improvement. Treatment of spinal cord injury also tested the use of intraspinal foetal central nervous system grafts. Three clinical studies have been performed where the results mainly focused on security of the procedure. However, encouraging results were obtained.

In particular translational correspondence was observed between preclinical locomotor performance in a cat before and after intraspinal foetal transplant surgery compared with a human subject before and after human foetal cell grafting. Transplantation of foetal central nervous system tissue is considered actually as a gold standard in neurobiology. Beside neurology, human foetal liver cells have been used already more than 10 years for transplantation to treat severe immunodeficiencies, haematological disorders and inborn errors of metabolism when there was no perfectly matched donor for marrow transplantation. Recently, human foetal skin cells derived from one cell bank (1–4 cm^2 tissue results in over 10.5 million foetal skin constructs) were used in clinical trials and new advances in tissue therapy are possible with cellular constructs obtained from ex vivo cultures. Engineered regeneration of human skeletal adult tissues could be also developed using human foetal bone cells. Surprisingly, tissue engineering using foetal cells has barely begun to be investigated. To evaluate their potential integration in a bone engineering strategy and following the experience gained in neurobiology, a biological characterisation of these cells is necessary, especially by evaluating the potential of foetal cells to produce mineralised bone extracellular matrix. Lately, a study was performed to specifically study the characteristics of human primary foetal bone cells for a better comprehension of their biology in vitro and to evaluate their potential use for tissue engineering in comparison to adult bone cells and mesenchymal stem cells. Human primary foetal bone cells were compared to adult bone cells and mesenchymal stem cells for their ability to proliferate and to differentiate into osteoblasts in vitro. Cell proliferation, gene expression of bone markers, alkaline phosphatase (ALP) activity and mineralisation were analysed during a time-course study. Human primary foetal bone cells were compared to osteoblasts and mesenchymal stem cells obtained from adult tissues for their ability to proliferate and to differentiate into osteoblasts in vitro.

The doubling time of foetal bone cells was comparable to mesenchymal stem cells but significantly shorter than for adult osteoblasts. Gene expression of cbfa-1, ALP, a1 chain of type I collagen and osteocalcin were upregulated in foetal bone cells after 12 days of treatment with osteogenic factors, showing higher inductions than for adult osteoblasts and mesenchymal stem cells. The increase of ALP enzymatic activity was stronger for foetal than for adult osteoblasts reaching a maximum at day 10, but lower than for mesenchymal stem cells. Importantly, the mineralisation process of bone foetal cells started earlier than adult osteoblasts and mesenchymal stem cells. The human primary foetal bone cells have the advantages of high proliferation rate as for mesenchymal stem cells and effective production of mineralised extracellular bone matrix as for adult osteoblasts. Human primary foetal bone cells represent an interesting and promising potential for therapeutic use in the bone tissue engineering field as these cells can be easily stocked "frozen for use" when necessary.

TISSUE ENGINEERING IN CARDIOVASCULAR SURGERY

For some years, one branch of experimental research at the DHZB has been the fascinating area of the artificial production of complex tissues, known as tissue engineering.

Engineering of Heart Valve Prostheses using Autologous Cells

For heart valve replacement, at present mechanical and biological prostheses are used. These function well but consist of material that is foreign to the body and have the disadvantage that they cannot grow in size. In children and teenagers this is a serious disadvantage as they "grow out of" the replacement valve and need several corrective operations. A potential solution to this problem may lie in the tissue engineering of replacement cardiovascular tissue.

The principle of tissue engineering is to use autologous cells (cells from the patient's own body) to produce viable, functioning replacement tissue that would become integrated into the surrounding tissue, grow with it and form highly durable valves.

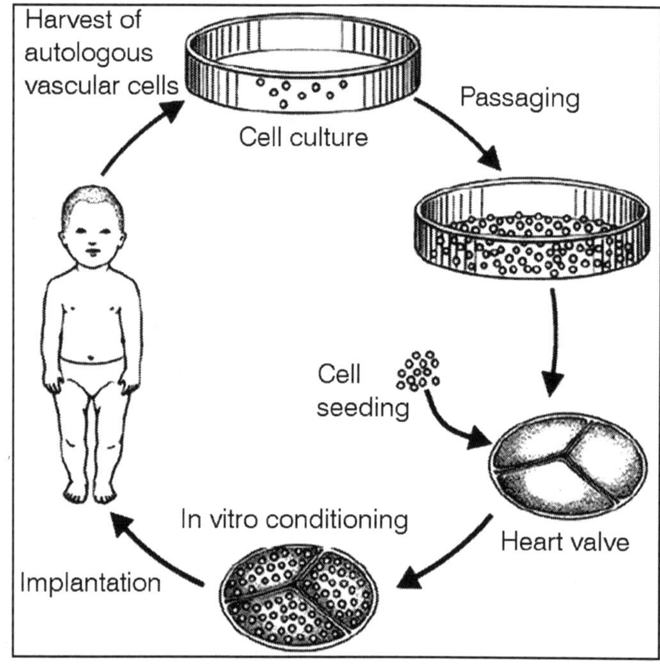

Fig. The principle of tissue engineering

The constructed valve fabricated from human cells and scaffold material, for example a biodegradable polymer, must be produced under optimal in vitro conditions so that it can be later be transplanted into the heart of the patient from whom the cells were taken. A particular focus of research of the DHZB laboratory is the identification of a source of suitable cells. Since it is not possible to use cells from heart valves themselves, cells must be utilized that are as similar as possible to heart valve cells. In the past few years human vascular umbilical cord cells have been evaluated as a source of autologous cells for cardiovascular tissue engineering. The umbilical cord is normally discarded as superfluous tissue after birth. It has several long blood vessels and can be used to harvest a very large number of vascular cells without the patient undergoing any harvesting procedure. These cells can be stored in liquid nitrogen until required.

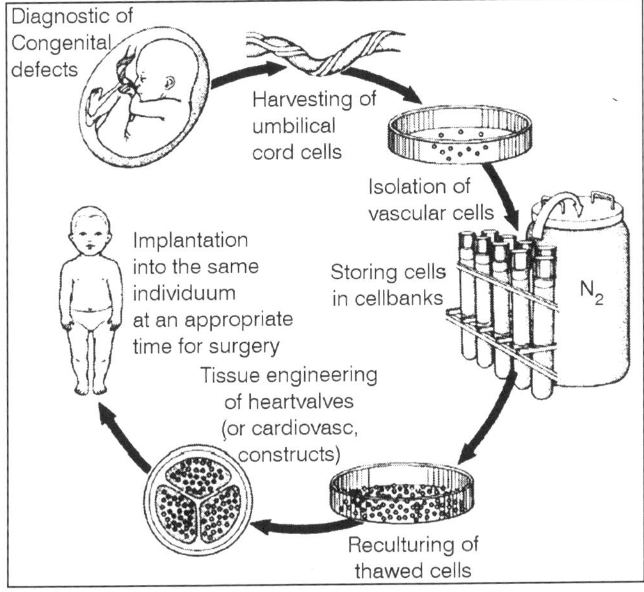

Fig. Cryopreservation of Vascular Umbilical Cord Cells

In our laboratory the vascular umbilical cord cells are characterized using immunohistological and molecular biological methods. The main target group for the possible use of these cells is newborn children with congenital heart valve defects. In older children and teenagers other suitable cells (mesenchymal precursor cells) are to be isolated from the patients' own bone marrow and used to engineer heart valves. A further focus of our research laboratory is the development of cell culture systems and "bioreactors" in which the heart valve construct is conditioned by mechanical means. By rotating the reactor, for example, it is possible for the cells to be evenly distributed on the scaffold material.

Fig. Cell-seeding Device

(1) rotating machine; (2) base frame with rubber rollers; (3) acrylic bowl with balance weight; (4) cell suspension inlet; (5) lid of the bowl; (6) cylindrical cell-seeding chamber.

These are the preconditions for the formation of tissue-like united cell structures on a complex three-dimensional scaffold like that for a heart valve. Influenced by pressure and flow in the bioreactor, the cells align in a particular direction. Thus the formation of extracellular matrix proteins is encouraged, which is very important for the development of cell tissue.

Fig. Bioreactor test

In parallel with this research, the in vitro conditions for fabricating tissue engineered valves are constantly being modified and optimized. The aim is to provide the cells with the greatest possible nutrient supply so that the growth process can be accelerated. In future, viable human heart valves consisting of autologous cells, with the ability to grow, are to be applied in cardiac surgery.

PREPARATION OF THE PVA-BIOACTIVE GLASS HYBRID FOAMS

The hybrid compositions prepared were 20, 30, 40 wt. (per cent) glass and 80, 70, 60 wt. (per cent) polymer. The hybrids were obtained using a procedure similar to the one described by in our previous work. An appropriate amount of the starting sol was added to the PVA solution. To this resulting solution were added the surfactant, sodium laureate sulfate (LESS), and HF 10 per centv/v solution. The mixture was foamed by vigorous agitation. Hydrofluoric acid was used to catalyze the gelation. The foams were cast just before gelation in plastic containers and sealed. The samples were, basically, aged at 40°C for 3 days and then dried at 40°C for 7 days.

HYBRIDS CHARACTERIZATION

FTIR was used to characterize the presence of chemical groups. The FTIR absorption spectra were obtained within the range between 4000 and 400 cm-1 (Perkin Elmer, Paragon 1000), using the KBr pellet method. For the PVA sample was prepared a thin film and used the transmission spectroscopy. Porous morphologies of the hybrids were examined with a scanning electron microscope (JEOL, JSM-6360L) after carbon coating.

The examination was carried out at accelerating voltage of 15 kV. To expose the internal architecture, the samples were cut with a razor blade. Hybrid PVA/bioglass foams were obtained for polymer contents of 60, 70 and 80 wt. (per cent). The use of two degree of hydrolysis of PVA changed the solution behaviour.

For PVA, fully hydrolyzed (Celvol 103), the concentrations of the starting PVA solution that proved to be adequate to allow the production and gelation of the foam were 23 and 28 wt. (per cent). Using 23 wt. (per cent), the resulting final foams have less residual water and the drying procedure is more efficient, compared to lower concentration PVA solutions used in previous work. Comparing with the 28 wt. (per cent) concentration, the amount of hydrofluoric acid necessary to gel the foam was higher.

Reaching gelation can be better controlled when working with PVA fully hydrolyzed, Celvol 103, since its solution has higher viscosity and the higher predominance of hydroxyl groups may contribute to crosslink with the polysiloxane network, through hydrogen bonds and condensation with silanol groups. For PVA, 80 per cent hydrolyzed, the starting PVA concentration of 23 wt. (per cent) did not work, even for higher HF concentration used foams could not be stabilized, since the gel-sol transition could not be reached using the procedure adopted.

At 28 wt. (per cent), depending of the hybrid composition, it was possible obtain the hybrid foams. For hybrid composition of PVA higher than 70 wt. (per cent), the control of the foam casting could not be done, since the gel-sol transition is a very slow process in the conditions used. PVA, 80 per cent hydrolyzed, presents higher amount of acetate groups left in the backbone, providing less reaction sites for reaction with silanol groups formed during hydrolysis of silicon precursors.

The presence of acetate groups in partially hydrolyzed polymers weakens the hydrogen bonding. This suggests that the mechanism involving interaction between the hydroxyl groups of PVA and hydroxyl groups of silanol, from hydrolyzed silica, is delayed. Chemical crosslinking, is a versatile method to modify and improve properties in materials.

In this way, there are several advantages on using organic-inorganic hybrid, because one may be able to tailor physical, biological and mechanical properties as well as predictable degradation behaviour where novel materials can be produced combining bioresorbable polymers and bioactive inorganic phases.

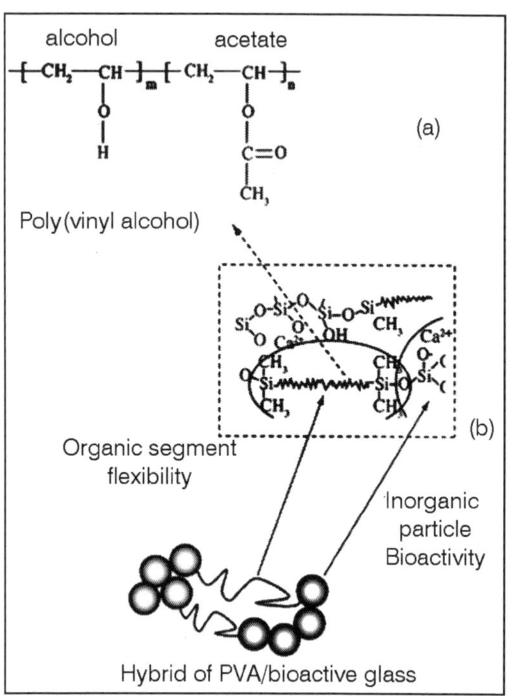

Fig. Schematic Drawing of the Inorganic -Organic Hybrid synthesized Based on PVA Polymer and Bioactive Glasses; a) PVA Chain with Variable Hydrolysis Degree; Hybrid Network Structure

FTIR Characterization

Representative IR spectra of the pure PVA films, the bioglass and the hybrid foams, for the sample group of PVA fully hydrolyzed, and the PVA 80 per cent hydrolyzed sample group. The spectra of the other hybrids obtained in this work are very similar to the spectra exhibited. No relevant differences were found by using the two PVA precursors neither for varying the PVA solution and composition of the hybrids.

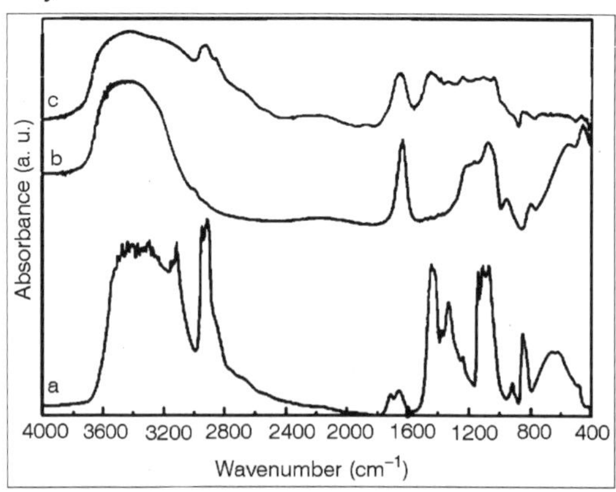

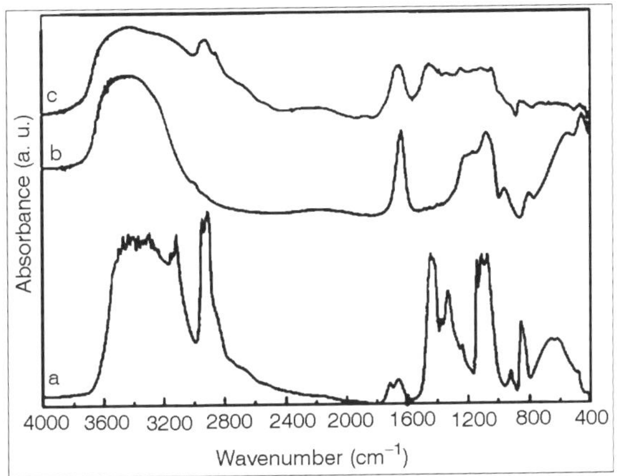

Fig. Ftir Spectra of Samples: a) Polyvinyl Alcohol, Fully Hydrolyzed, Celvol 103; b) Bioglass; c) Hybrid 70 wt. (per cent) PVA -30 wt. (per cent) Bioglass (3B-7Pc).

FTIR spectra of pure PVA are shown. For both PVA spectra, the broad band observed from 3100 to 3600 cm-1 may be assigned to O-H stretching due the strong hydrogen bond of intramolecular and intermolecular type. The presence of a higher hydroxyl content in the fully hydrolyzed PVA compared to the PVA 80 per cent hydrolyzed can be seen from the broader band around 3300 cm^{-1}. The C-H alkyl stretching band can be observed at 2850-2950 cm^{-1}.

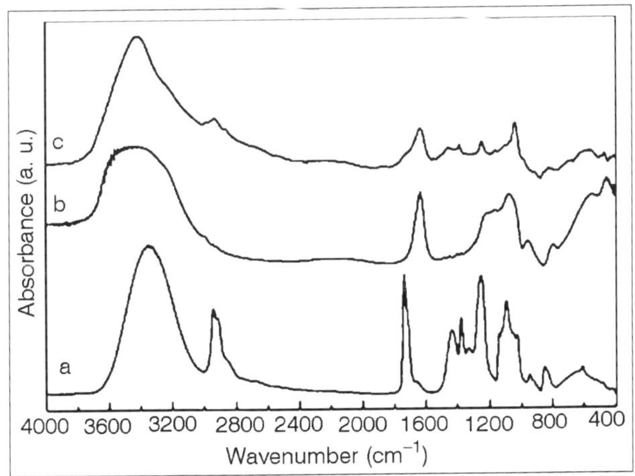

Fig. Ftir Spectra of Samples: a) Polyvinyl alcohol, 80 per cent Hydrolyzed; b) Bioglass; c) Hybrid 70 wt. (per cent) PVA -30 wt. (per cent) Bioglass (3B-7Ph)

The absorption peaks at approximately 1710 cm^{-1} (for fully hydrolyzed PVA) or 1740 cm-1 (for 80 per cent hydrolyzed) and at 1090-1150 cm^{-1} may be attributed to the stretching vibration of C = O and C-O of the remaining vinyl acetate non-hydrolyzed group of PVA polymer. The absorption band around 1700 cm^{-1} arises due to the carbonyl band (C= O) of the acetate group found in partially hydrolyzed PVA polymer and has higher relative intensity for the PVA 80 per cent hydrolyzed as expected from the higher concentration of remaining acetate groups.

In the FTIR spectra of the bioglass presented in both figures, the vibrational modes due to Si-O-Si asymmetric stretching are observed at approximately 1070 cm-1 and 1200 cm-1. Additional peaks are seen at 790 cm^{-1} (for symmetric Si-O-Si stretching vibration, and at 450 cm^{-1} (for Si-O-Si bending modes). A typical absorption band observed in silica gel is located at 1640 cm^{-1} and is attributed to the deformation mode of adsorbed molecular water in the pores. The band at 950 cm^{-1} has been ascribed to the stretching vibration of silanol groups in pure silica (vibrational modes of Si-O- stretching).

FTIR spectra of hybrid in the composition 70 wt. (per cent) PVA30 wt. (per cent) bioglass are presented for Celvol and 80 per cent hydrolyzed, respectively. The bioglass component is identified by major vibration bands, (Si-O-Si, at 1080 and 450 cm^{-1}). For both hybrids, the peak at 950 cm^{-1} associated with the Si-OH vibrational mode remains as a shoulder. In the frequency range of 3000-3650 cm^{-1}, a broader band was noted, especially for hybrid prepared from PVA fully hydrolyzed.

The band located at 1640 cm-1 remains in the spectra of the hybrids since the low heating temperature is not enough to remove the molecular water of the pores. The terminal vinyl groups of the PVA which appear at 1710 cm^{-1} (for fully hydrolyzed PVA) or 1740 cm^{-1} (for 80 per cent hydrolyzed) are not shown in the hybrids, which suggests that these bonds might be involved in crosslinking with silica. In the range 1500-900 cm^{-1} there is a superposition of the bands derived from the bioglass and the PVA components.

Scanning Electron Microscopy

The pore morphology and distribution of the hybrid foams can be seen by the SEM micrographs presented in Figure. Figure shows SEM micrograph of hybrid 70 wt. (per cent) PVA-30 wt. (per cent) Bioglass (3B-7Pc), SEM micrograph of hybrid 80 wt. (per cent) PVA-20 wt. (per cent) Bioglass (2B-8Pc), Figure shows SEM micrograph of hybrid 70 wt. (per cent) PVA-30 wt. (per cent) Bioglass (3B-7Ph 35), and Figure shows the micrograph of 60 wt. (per cent) PVA-40 wt. (per cent) Bioglass (4B-6Ph-35).

Very distinct features can be observed in the different hybrids produced. The observed morphologies of the foams vary considering the processing variables. The porosity and pore sizes are strongly dependent on the hybrid composition, concentration of the polymer solution used and gelation time. The control of the gelation time is very important to establish an adequate morphology of the hybrids produced by this method. For composition of 70 wt. (per cent) PVA, for both PVA polymers used, the gelation time was found to be approximately 13 minutes and this is a proper time to gel the three-dimensional hybrids structure and stabilize the bubbles formed during the process. These hybrids exhibited macroporous structure with interconnected open pores and pore size varied from 50 to 350 mm. For the composition of 60 wt. (per cent) PVA, the gelation time was shorter and the time at which the foam was cast, the gelation point almost reached, affecting the final porous structure.

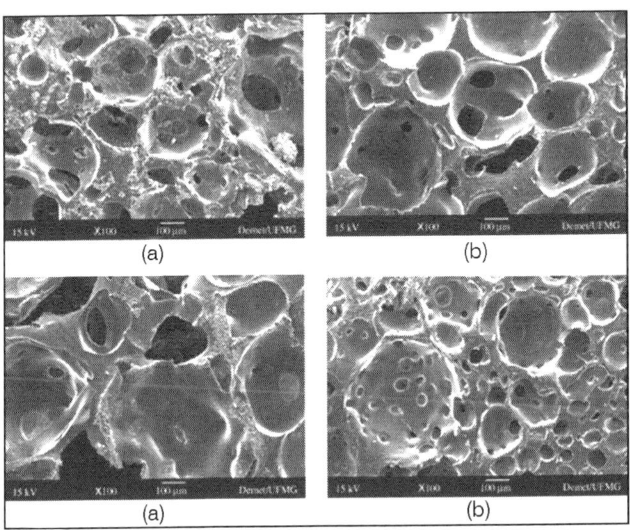

Fig. Scanning Micrographs of Hybrids: a) 70 wt. (per cent) PVA -30 wt.(per cent) Bioglass (3B-7Pc); b) 80wt. (per cent) PVA -20wt. (per cent) Bioglass (2B-8Pc); c) 70wt. (per cent) PVA -30wt. (per cent) Bioglass (3B-7Ph-35); d) 60wt. (per cent) PVA -40wt. per cent) Bioglass (4B-6Ph-35)

For this hybrid, the distribution of pores is more homogeneous and pore size ranged from 50 to 90 mm and reaching 450 mm for the bubble-like pores. For the composition of 80 wt. (per cent) PVA, For both PVA polymers used, the gelation time was delayed and the pore morphology was semi-spherical. Hybrid showed in Figure has the largest pores (pore size from about 160 to 500 mm).

The number of possible combinations of bioactive glasses with biocompatible polymers such as PVA is almost countless. That means each and every proportion of inorganic phase CaO, SiO_2 and P_2O_5, not to mention the synthesis parameters themselves, will result on a different hybrid system with some specific properties and bioactive behaviour. Therefore, synthetic bioactive and bioresorbable composite materials are becoming increasingly important as scaffolds for tissue engineering. Next generation biomaterials must combine bioactive and bioresorbable properties to activate in vivo mechanisms of tissue regeneration, stimulating the body to heal itself and to facilitate replacement of the scaffold by the regenerating tissue.

Bioactive glass reacts with physiological fluids to form tenacious bonds to hard (and in some cases soft) tissue. However bioactive glass is a stiff and brittle material, difficult to form into complex shapes and prone to catastrophic fracture under loads. Conversely, synthetic bioresorbable polymers are easily fabricated into complex structures, yet they are too weak to meet the demands of surgery and the in vivo physiological environment. Composites of tailored physical, biological and mechanical properties as well as predictable degradation behaviour can be produced combining bioresorbable polymers and bioactive inorganic phases.

Hybrid Polyvinyl alcohol/bioglass foams have been prepared by the sol-gel method, with a high content of polymer (60 to 80 per cent). The synthesis parameters need to be very well combined in order to allow the control of foaming and gelation. Pore characteristics of the obtained scaffolds have been related to hybrid composition, concentration of the polymer solution used, and gelation time. The macropore diameter ranged from 50 to 600 mm. The results suggest that there is scope for designing materials with controlled compositions and morphologies. Such system has a potential to be used in bone tissue engineering, but further studies must be conducted.

THE TISSUE ENGINEERS AS A GROUP

Disciplinary affiliations of tissue engineers are difficult to analyse in a precise quantitative way, because the specific mix of research areas that are included in a department with a given name or are associated with a degree with a given designation, and the number and character of departmental affiliations awarded to faculty, vary from one institution to another in idiosyncratic ways. Nevertheless, overall trends emerge clearly from a rough tally of the available data. Viewed in terms of both the departments by which their doctoral degrees were awarded and the departments in which those who are in academia now hold faculty appointments, tissue engineers are indeed predominantly engineers.

By training, more than half of the individuals for whom we have specific data are engineers. Approximately one fifth of the tissue engineers in our sample hold medical or dental degrees, often in conjunction with an engineering or science doctorate. In view of the critical importance of cells and of physiology in tissue engineering, the fraction of individuals who hold only a doctorate in biological sciences (for example, biology, biochemistry, or non-clinical medical sciences such as physiology or anatomy) is remarkably small – roughly one out of ten.

Chemical engineering is by a wide margin the engineering discipline most frequently encountered in this cohort, followed by biomedical or bioengineering, and by mechanical engineering. A more liberal interpretation of which aspects of contemporary orthopedics and biomechanics research should be considered part of tissue engineering would perhaps have yielded somewhat increased weights for biomedical and mechanical engineering in the sample. Current academic departmental affiliations of these tissue engineers are also strongly weighted toward engineering. However, here bioengineering or biomedical engineering is the leading disciplinary affiliation by a wide margin, followed by chemical engineering and, in a distant third place, mechanical engineering.

This pattern may reflect the relatively recent emergence of biomedical engineering as a discipline, and the migration of faculty – especially recent trainees taking their first faculty appointments – from some of the classical engineering disciplines to newer and often growing departments of biomedical engineering. Biological science affiliations are few, and are weighted toward basic medical science departments. Clinical and clinical science departmental affiliations are strongly weighted toward surgery and surgical specialties, notably orthopedics.

This latter pattern may to some extent be an artifact of selection bias in construction of the sample; for example, it is likely that a more liberal definition of tissue engineering to include a greater part of islet cell transplantation research and a correspondingly more aggressive search for clinicians who have been involved in such work would have resulted in the inclusion of more endocrinologists in the roster. Nevertheless, we would expect surgeons to dominate in any reasonably representative sample of clinician-scientists active in tissue engineering. At present, most of the individuals active in this representative sample of tissue engineers entered the field after completing dissertations in other areas. Because of the interdisciplinary character of tissue engineering, the field will likely continue to have a relatively high proportion of post-doctoral-training entrants. However, over time, the fraction of individuals whose thesis research was explicitly in tissue engineering, and the representation of universities other than MIT with broad platforms of research and training activity in tissue engineering, such as Rice, UCSD and Georgia Tech, should increase as a proportion of any roster of researchers active in TE.

One last characteristic of this cohort is noteworthy here. The great majority of individuals are involved in TE on only a part-time basis. That is, academic tissue engineers typically maintain a number of lines of research, some of which meet any reasonable definition of tissue engineering, some of which straddle the ill-defined boundary that delineates the field, and some of which are clearly situated within other intellectual domains. By this measure, tissue engineering could be viewed more as a tactic than as a discipline – as part of an interdisciplinary assault on unsolved therapeutic challenges that draws opportunistically on an ever-wider range of knowledge and tools from science and engineering.

TISSUE ENGINEERING AND LEADING CENTERS IN THE FIELD

A bibliometric analysis of papers important to the field of tissue engineering revealed a list of more than 350 US institutions active in research. Our interviews with experts in the field as well as extensive secondary research validated this list and also pointed toward several other institutions as being highly influential in the field—not only in terms of their research contributions, but also in their training of personnel currently active in the field.

Boston Area: MIT and Harvard

Perhaps the single most prominent geographical and institutional locus of research in tissue engineering has been in the Boston area, centered on the Massachusetts Institute of Technology (MIT) in Cambridge and the Harvard Medical School (HMS) in Boston, although researchers at other Boston-area institutions have been involved as well. MIT was the site of three independent, seminal lines of research on substitutes for human skin. During a period from the mid-1970s through the mid-1980s, the laboratory of cell biologist Howard Green, MD achieved a breakthrough in the cultivation of human keratinocytes, developed it into a method for growing epithelial grafts from a small piece of autologous epidermis, and demonstrated the viability of this method in the treatment of burn victims.

In 1987, this technique became the basis for a startup company called BioSurface Technology that offered cultured epidermal autografts commercially. The company was acquired by Genzyme in 1994 and became Genzyme Tissue Repair, and the service is still offered under the name Epicel. Dr. Green himself has continued his career as a cell biologist at HMS, but has not been prominent in the subsequent development of tissue engineering. Ioannis Yannas joined the faculty in the mechanical engineering department at MIT in 1966, after completing his PhD at Princeton. From the start, his research focused on the properties of collagen and its role in connective tissues, using approaches from chemistry, physics, and biomechanics. By the 1970s, Yannas was investigating the use of acellular collagen-glycosaminoglycan matrices as wound dressings designed to serve as biodegradable templates for the regeneration of viable skin.

In 1977, Yannas, in collaboration with Massachusetts General Hospital and Shriners Burns Institute surgeon John F. Burke, was awarded a patent for a "multilayer membrane useful as synthetic skin". In 1980, they published the first in a series of papers outlining design considerations for what they called an "artificial skin", and shortly thereafter, they reported the successful use of such an artificial skin in the treatment of extensive burn injury.This research resulted in the development of a commercial product, Integra Dermal Regeneration Template, which is currently licensed to and manufactured by

Integra Life Sciences Corporation. Prof. Yannas has continued to maintain an active research programme on the principles and applications of induced tissue regeneration. Eugene Bell, a long-time faculty member in the biology department at MIT, also pursued in the 1970s a line of research that examined interactions between collagen and skin cells. In 1979, Bell and colleagues published a paper describing the production in vitro of a tissue-like structure, obtained via contraction of a collagen lattice seeded with fibroblasts. In 1981 Bell's group reported the successful grafting of a "living skin equivalent" consisting of fibroblasts cast in collagen lattices and seeded with epidermal cells.

A patent was obtained for this "tissue-equivalent" in 1984, and Bell left MIT in 1986 to found Organogenesis, Inc. and pursue commercial development of the product. He left Organogenesis in 1991 to return briefly to MIT, but left again in 1992 to found Tissue Engineering Inc. Bell has led an active proprietary research programme at TEI Biosciences, focusing on the further refinement of collagen fibre-based scaffolds and the enrichment of these scaffolds with signaling molecules that induce stem cell development and tissue regeneration, but has published little in the peer-reviewed biomedical literature in recent years.

Langer's graduate studies were carried out under the supervision of yet another MIT researcher active during the "prehistory" of TE, Clark Colton of the chemical engineering department. Colton's own PhD, completed in the same department in 1969 under the supervision of Kenneth A. Smith, was entitled Permeability and Transport Studies in Batch and Flow Dialyzers with Applications to Hemodialysis. Colton's research on filtration technologies relevant to artificial organs continued following the completion of his PhD, and he collaborated with William Chick at Harvard Medical School, Pierre Galletti at Brown University and colleagues on the research which led to a 1975 publication in the Transactions of the American Society for Artificial Internal Organs on a "hybrid artificial pancreas" for rats consisting of beta cells cultured on synthetic semipermeable hollow fibers. Colton was the only senior MIT researcher represented with a paper at the Granlibakken workshop in 1988.

Although contemporaneous with this early work on artificial organ technologies in Colton's lab, the research that formed the basis for Langer's 1974 ScD in chemical engineering was part of a separate line of work being pursued by Colton in the area of enzyme engineering. Langer proceeded to a postdoctoral fellowship in the laboratory of Dr. Judah Folkman at Harvard Medical School and Children's Hospital in Boston. Folkman's pathbreaking work on implantable drug delivery systems and on angiogenesis in cancer addressed a number of fundamental issues that were to be of broader significance for tissue engineering, including identification and analysis of the factors that govern the growth and differentiation of blood vessels, and the design of delivery systems to enable the sustained release of bioactive proteins and other molecules in a desired location in order to influence tissue growth. Langer's entry into tissue engineering was catalyzed by a collaboration with Jay Vacanti. Vacanti earned his MD at the University of Nebraska, and came to Boston in 1974 to begin his postgraduate training in surgery.

He completed a residency in general surgery at the Massachusetts General Hospital during 1974- 1981, followed by a fellowship in pediatric surgery at Children's Hospital from 1981-1983. During the period of his residency at MGH, he spent two years – from 1977 to 1979 – in the laboratory of Dr. Folkman, where his collaboration with Robert Langer began.

Their first joint publication, written in collaboration with Folkman and two other colleagues, reported on experiments in the inhibition of tumor growth in animal models by regional infusion of a cartilage-derived angiogenesis inhibitor. Returning to MIT following his postdoctoral work with Folkman, Robert Langer built a highly productive research and training programme focused primarily on polymer-based drug delivery systems and their applications.

Vacanti, following his pediatric surgery fellowship, received two further years of clinical training in the renowned transplantation programme at the University of Pittsburgh, then returned to Boston in 1985 to join the staff at Children's Hospital and launch his own clinical and research programme. Vacanti's clinical experience in transplantation in Pittsburgh, and then in launching a pediatric liver transplantation programme in Boston, confronted him with the intractable problem of organ supply. Following his return to Children's Hospital in Boston in 1985, he began to explore with Langer the problem of how to extend to three dimensions the early success of Eugene Bell in growing flat sheets of tissue from skin cells.

From this emerged the concept of using a porous, resorbable, three-dimensional polymer scaffold as a template for cell growth, and the 1988 paper the Journal of Pediatric Surgery reporting its experimental application. This technique was to be enormously influential in shaping the work of other investigators who entered the field – indeed, it could be said that it provided the intellectual "scaffolding" for the work of an entire cadre of investigators who shared the long-term vision of creating "replacement" solid organs for therapeutic purposes.

Although Langer himself is not primarily a tissue engineer, TE projects have remained an active line of research in his laboratory. He has mentored an extraordinary number of doctoral and postdoctoral trainees, some of whom focused their work specifically on tissue engineering applications. Many have gone on to establish independent academic careers in which tissue engineering research has played an important part.

INSTITUTIONAL SUPPORT OF TISSUE ENGINEERING

In addition to the individual researchers in academia and industry, key players in a field can include the federal agencies that provide financial support, professional societies and other institutions designed to promote and advance a field's mission. Professional societies may promote educational programs, research applications, and the development of professional standards in an overall effort to advance the dissemination of knowledge in a particular field or research area.

As networking facilitators, conferences, meetings, and other symposia supported by societies may be instrumental in bringing together researchers who work on similar topics but may not otherwise collaborate. Several groups have emerged in tissue engineering's development.The National Science Foundation is one of many organizations – including the U.S. Federal government and foreign governments, industry and non-profit foundations – that have funded tissue engineering research in the United States and around the world. The WTEC report estimates that nearly $3.5 billion dollars have been spent on tissue engineering in the last decade. Of this, less than 10 per cent has come from the U.S. Federal government.

According to their definition, the National Science Foundation, primarily through its Directorate for Engineering, provided less than 1 per cent of worldwide support for tissue engineering, but nearly 7 per cent of U.S. Federal government support. The accompanying examine U.S. Federal government funding of tissue engineering in greater detail, using a more restrictive definition. They provide a partial glimpse into Federal support of TE using data from RaDiUS, a database that tracks the research and development activities and resources of the Federal government. Of all Federal agencies supporting biomedical science and engineering, NSF is second only to NIH in providing financial support to TE, and has increased both its level and share of Federal funding in the past few years. Before turning to NSF's role in supporting tissue engineering, we provide brief descriptions of major efforts of other Federal agencies that have been active in tissue engineering in different ways.

National Institute for Standards and Technology (NIST)

NIST's Advanced Technology Programme (ATP) began in 1988 and was designed to focus on commercial applications of high-risk cutting edge basic scientific research, such as biology, a field NIST did not yet participate in. The first ATP competition was held in 1990-91 and the first TE award was made in the 2nd round of competition to Acerm Biosciences, a company growing stems cells from bone marrow. In the years subsequent, NIST added projects at the rate of 1-2 per year on topics such as building better scaffolding, bioreactors, and xenotransplantation.

Serious efforts in tissue engineering, however, did not occur until the mid-late 1990s. In 1993-94, a two day conference was held on cutting-edge biotechnology and involved individuals from several agencies including Francis Collins of NIST, Fred Heineken of NSF, and Kiki Hellman of FDA. The group identified 'hot topics' in a hierarchy—tissue engineering being one of those near the top of the list. For each, a set of key issues was identified and listed: in what areas is there enough basic science research and where is there a black hole? Other technologies identified at this meeting included DNA diagnostics, TE, vaccines, gene therapy, and biosensors. Approaches for the development of specific industries were discussed. Stan Abramowitz, who headed the early NIST/ATP efforts, was heavily involved in this. Staff attended conferences to "find the right people" to get advice, and to head these initiatives. As a follow-up to these efforts, a series of white papers were solicited in 1995 in preparation for a focused competition in TE. The first focused programme was initiated in 1997 when 56 proposals were submitted and 12 awarded (for a total of $12 million that year).

Focused competitions in tissue engineering ended for ATP in 1998, however. The ATP Programme has been instrumental in funding high-risk commercial applications of ideas that venture capitalists are unlikely to support—typically these are multidisciplinary projects which require academic collaboration. Dr. Rosemarie Hunziker, a former Programme Officer at ATP, believes that NIST/ATP support of TE gave it credibility, an important action, especially in light of its heavy private sector orientation.

National Aeronautics and Space Administration (NASA)

Beginning in the late 1970's, NASA's efforts in tissue engineering have focused on development of techniques for three-dimensional cell and tissue culture. The agency's interest in tissue engineering comes from a need to understand how cells and tissue behave in space. Laboratory cultures which mimic live human tissue can be used as models to conduct space research and research on deadly diseases, such as cancer—eliminating the need for human test subjects. In the late 1980's, the work of Milbourne and Wolf of the Johnson Space Centre culminated in the development of the rotating bioreactor, a horizontal rotating wall vessel with a centre oxygen membrane, which promoted cell and tissue growth under a variety of conditions. A patent on the device was filed in 1988 and issued in 1991. While initially referred to under the agency's biotechnology heading, NASA officially recognized these efforts as "tissue engineering" when the term appeared for the first time in a 1994 programme update. Since then, NASA has continued to fund selective research on bioreactors and related topics and has provided a steady stream of funding to individuals in this area. Much of the work done by Lisa Freed and Gordana Vunjak-Novakovic of MIT has been supported by NASA's Cellular Biotechnology Area.

Food and Drug Administration (FDA)

The FDA has taken an interest in tissue engineering since the early 1990's. In 1990, the Centre for Devices and Radiological Health (CDRH) in conjunction with the State of Maryland held a conference to examine biological applications with device use. Among the topics raised was tissue engineering. Dr. Kiki Hellman, head of the CDRH, saw potential in the field and became interested in exploring it further.

Two years later, a workshop was held to examine promising new technologies—tissue engineering being one of them. From an agency perspective, the FDA was proactively involved in the technology's development. According to Dr. Hellman, the FDA made a concerted effort to deal with in-house needs (staff training, education) so that they could anticipate the technology and be prepared to deal with the any technical or scientific issues that would arise. Many of our interviewees also state that the FDA has been instrumental in allowing products to move to market, encouraging the development of standards to measure devices (in conjunction with the ASTM Committee F04 Medical and Surgical Materials and Devices). In all, many experts in the field laud their efforts, stating that the FDA is not a force standing in the way of progress.

The Tissue Engineering Society International (TESI)

In 1995, Charles Vacanti started The Tissue Engineering Society. In its early days, the society served to foster communication between individuals in the tissue engineering world through sponsorship of meetings and conferences. In 1995, the Society published the first issue of Tissue Engineering, the first dedicated journal in the field.

Despite the placement of several prominent tissue engineers on its editorial board, the journal lacked popularity. Many researchers continue to preferentially publish in more traditional disciplinary journals or general peer-reviewed publication like Science or Nature. More recently, the society has grown to include a more international focus and serves as a networking organization to foster dialogue between international researchers in the field.

The Pittsburgh Tissue Engineering Initiative

A unique and somewhat unusual contributor to the field of tissue engineering has been the Pittsburgh Tissue Engineering Initiative (PTEI). The organization was founded by Peter Johnson a former surgeon at the University of Pittsburgh. After becoming chairman of the department of plastic surgery, Johnson was in a unique position to form a network of scientists in the Pittsburgh area. Based largely upon his own interest in tissue engineering and its potential, Johnson pushed to develop a joint technology transfer policy between the major Pittsburgh area universities, and PTEI was born. In its beginnings, PTEI served mostly as a networking mechanism—to raise funds to provide support to TE research in the surrounding area. The initiative had four key components:

- Technology development grants,
- Summer research internships to increase student base in TE,
- Biotech exposure (with a minority focus),
- A guest speaker programme (a coordinated effort to bring big names in the field to Pittsburgh).

PTEI raised money and support and led the 1st TE biotech company in the area. Currently, there are at eight firms in the area. PTEI has encouraged an amalgamation and growth of the critical mass necessary to sustain development in TE for the Pittsburgh community. As a result, Pittsburgh is rapidly becoming a centre for informatics—a new branch in the field of tissue engineering. PTEI also serves as a model for other cities to advance research and development in TE. Toronto is expanding their TE workforce utilizing the PTEI model.

The Whitaker Institute for Biomedical Engineering

The mission of The Whitaker Foundation is to promote better human health through advancements in medicine. This is accomplished through a series of competitive grant programs that support research and education in biomedical engineering at academic institutions in the United States and Canada. Whitaker has been instrumental in funding bioengineering programs around the country. Most notable for TE, they, in 1999, established the Whitaker Institute for Biomedical Engineering (WIBE) at UCSD under the leadership of Y.C. Fung. Much of the fundamental work supported under this institute will contribute to advancements in the field of TE. However, the foundation has been reluctant to classify TE as an independent entity, but rather a sub-field of bioengineering.

Multi-Agency Tissue Engineering Sciences (MATES) Working Group

More recently, many agencies of the federal government have become interested in supporting the mission of tissue engineering. In an effort to coordinate this support, representatives from each of several federal agencies came together in 2000, under the leadership of Drs. Fred Heineken (NSF) and Kiki Hellman (FDA) to form the Multi-Agency Tissue Engineering Science (MATES) Working Group. MATES has three major goals:

- To facilitate communication (and prevent redundancy of funding) across departments/agencies by regular information exchanges and a common web site,
- Enhance cooperation through co-sponsorship of scientific meetings and workshops, and facilitate the development of standards,

- To monitor technology by undertaking cooperative assessments of the status of the field.

MATES has developed a web-site which they hope will become "onestop-shopping" for those seeking information on the field (such as information about federal funding, scientific meetings, regulatory guidance and standards development). Participating agencies in MATES include the Department of Commerce (NIST), the Department of Energy (DOE), the Department of Defence (DARPA), the Department of Health and Human Services (FDA, NIH), the National Aeronautics and Space Administration (NASA), and the National Science Foundation (NSF).

The panel, co-chaired by Drs. Frederick Heineken of NSF and Kiki Hellman of the FDA also focuses on some regulatory, legal, and ethical issues for TE. A particular area of interest has been a global regulatory initiative for TE—an effort that will require the development of standards for testing tissue engineered products. A major contribution of the MATES group is the recent World Technology Evaluation Centre (WTEC) study on tissue engineering, which attempts to summarize the technical contributions of current work in TE and to estimate the funding for TE from each of the federal agencies. The study has become a focal point for activities of the working group, under the auspices of the Subcommittee on Biotechnology, Committee on Science of the President's National Science and Technology Council (NSTC).

The results of the WTEC study were used by MATES to plan a joint interagency programme announcement in tissue engineering issued under the new National Institute for Biomedical Imaging and Bioengineering, which solicits research aimed at addressing specific gaps in tissue engineering.

The Role of the National Science Foundation (NSF)

The National Science Foundation, particularly through its Directorate for Engineering, appears to have played an important role in the emergence of tissue engineering as a recognized field of activity, and in shaping the character and in determining the direction of the field.

Financial Support

From an examination of Fastlane Award data as an illustration of NSF support for tissue engineering, it is evident that the majority of NSF support goes toward individual investigator awards and Centers research. NSF is also a strong supporter of workshops, conferences and other meetings, which suggest a prominent role in support of networking activities among individuals active in the field. Analysis of NSF funding by division shows that the vast majority of the agency's support of tissue engineering—more than 80 per cent—emanates from the Bioengineering and Environmental Systems (BES) and Engineering, Education and Centers (EEC) Divisions—both of which reside under NSF's Directorate for Engineering.

Researcher Support

While the interviews did not shed much light on the role of NSF as a funder, bibliometric analysis revealed that NSF has had a significant role in supporting lead researchers in the field. The relative role of NSF support of TE over the years. NSF support has been acknowledged on 12 per cent of the papers revealed by the bibliometric analysis.

The influence of NSF funding on top researchers in the field, however, has been disproportionately large. Three of the most prolific researchers – Vacanti, Langer and Mooney - acknowledge NSF funding on 19 per cent-37 per cent of their papers, including on the 1993 Science paper (this percentage may, in fact, be even higher, as not all papers list acknowledgement of research funding and, further, some researchers fail to mention all of the funding sources that should be cited in acknowledgements).

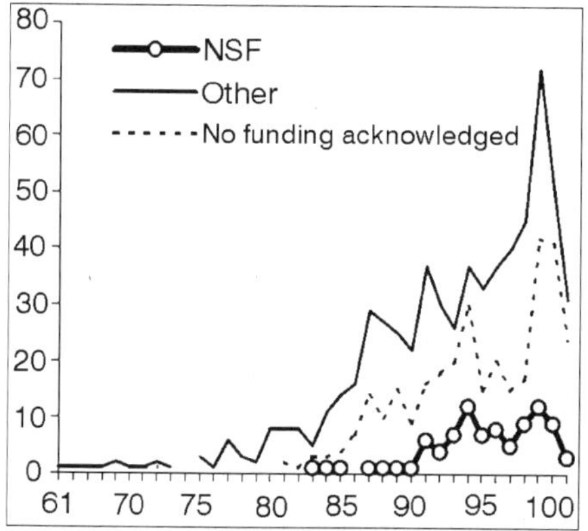

Given the wide range of tissue engineering activities, it is not surprising that NSF support of researchers varies considerably. Notable researchers such as Linda Griffith, Jeffrey Hubbell, P.M. Kaufmann, and Mark Saltzman also acknowledge NSF on a third or more of their papers, while NSF appears to have had no role in supporting other prominent researchers such as Arnold Caplan, Lisa Freed, V.M. Goldberg or Gordana Vunjak-Novakovic. NSF funding, however, is not evenly distributed across all aspects of tissue engineering.

NSF support is clearly targeted toward the application of engineering principles, especially with the aim of advancing engineering knowledge. The bibliometric study to identify the sub-fields of tissue engineering where NSF-funded researchers have been most active and demonstrates that the agency has held true to its stated mission and goals. Almost 90 per cent of NSF-supported papers are in fields related to engineering or fields that promote the application of engineering principles to solve problems in a variety of realms.

Institutional Development

NSF has played a small but significant role in developing institutions and networks in the field of Tissue Engineering.

Conferences and Workshops

NSF has been an early sponsor of workshops and conferences related to TE. The term "tissue engineering" itself was proposed at a 1987 panel meeting convened to consider future directions for the Engineering Directorates's Bioengineering and Research to Aid the Handicapped Programme. Strong interest in this concept within the Engineering Directorate led to a special panel meeting on tissue engineering, again at NSF, in the fall of 1987, and then to the Granlibakken workshop of 1988, now considered to be the first formal scientific meeting of the emerging field of tissue engineering.

In all, about one per cent of NSF's total TE support in the period 1987-2001 was devoted to networking and conference support, and many of the individual investigator awards in TE were funds to support presentations of results from research conducted with the NSF support and attendance at non-NSF conferences and meetings.

MATES

As a founder and critical participant in the MATES Working Group, the Engineering Directorate, through the leadership of Frederick Heineken, Kiki Hellman, and others, has made a pronounced effort in recent years to engage participation from a range of disciplines. The MATES group aims to enhance collaboration and cooperation amongst several federal agencies so that gaps in tissue engineering research can be addressed. An important example is an outgrowth of the MATES initiative, a study published through WTEC, which summarizes the global status of the field and notes where progress is needed.

NIBIB

Aside from its participation in MATES, NSF continues to collaborate with other Federal agencies and programs on TE related research and education. In particular, NSF has ongoing collaboration with the newest of the NIH Institutes, the National Institute for Biomedical Imaging and Bioengineering (NIBIB), established in 2000. The first "official" NIBIB interagency activity was a joint NIH/NSF Workshop on Bioengineering and Bioinformatics Education and Training.

Since then, the two agencies have established the Bioengineering and Bioinformatics Summer Institutes (BBSI) Programme to provide students majoring in the biological sciences, computer sciences, engineering, mathematics, and physical sciences with interdisciplinary bioengineering or bioinformatics research and education experiences137.Another important outgrowth of the NIBIB is a recent solicitation (dated December 30, 2002), likely influenced by the work of MATES and the WTEC report, seeking proposals to address key research challenges in TE138. The initiative is expected to draw biologists, clinicians, and engineers to work together to address issues such as cell sourcing, cell identification and characterization, engineering design, and other enabling technologies necessary to move the field forward. A total of $8 million dollars has been earmarked for the initiative.

Centre Support

NSF's Directorate for Engineering has provided significant support toward institutional development for TE through its funding of Centers. These include the Georgia Tech/Emory ERC, which was awarded in 1998, and the University of Washington Engineered Biomaterials (UWEB) Centre.

Aside from NSF, the bulk of institutional support in TE has been provided by the Whitaker Foundation, through its role in building the field of biomedical engineering, and especially in providing institutional development funds for the creation and expansion of bioengineering departments– not only at the academic centers highlighted, but at many other institutions across the nation.

These departments are increasingly taking over from departments of chemical and mechanical engineering as the focus of academic research activity in tissue engineering, and the drive to create and expand TE focus areas in many of these emerging departments has created opportunities for young investigators completing their training at the institutions that have led the way in TE research.

Not evident in these funding analyses, but worthy of note, is that NSF has been an important source of support over the years for the work of Douglas Lauffenburger, a highly productive researcher on the fringes of tissue engineering who has trained many younger investigators involved in the field. Since 1999, Lauffenburger has served as director of the MIT Biotechnology Process Engineering Centre (BPEC), an NSF-funded ERC, in its second incarnation in 1995. BPEC has developed new research thrusts that have moved the focus of its activity toward the boundary between tissue engineering and several adjacent fields, with projects led by several MIT investigators prominent in tissue engineering.

While available evidence suggests that no single organization engendered monumental changes in the field of tissue engineering, NSF's Directorate for Engineering did create an environment that encouraged the crystallization of an emerging concept, both with its innovative Engineering Research Centers (ERC) Programme and its ongoing internal efforts to define productive future directions for its bioengineering programme. NSF hosted the first meetings at which the concept of tissue engineering as a focus of research was defined and strategies for its future growth and potential discussed. While the

Directorate for Engineering has no doubt provided funding support to key researchers in the field, it is more difficult to attribute specific research contributions and breakthroughs to the Directorate.

In fact, in general, the broad and interdisciplinary nature of the field make it difficult to classify any number of research contributions as seminal. Efforts to "engineer" tissues to restore structure or function predate the events of 1987-88 by many years.

However, if there is a research development that marked the emergence of a recognizably distinct field on this foundation of many related, pre-existing lines of research, it appears that the consensus in the research community and through our own best judgment points to the January 1988 publication by Langer, Vacanti and colleagues describing the strategy of using resorbable artificial polymer matrices seeded with cells as a vehicle for cell transplantation.

This early phase of the Langer/Vacanti collaboration, catalyzed by the shared experience of working under Folkman, brought together at least four of the conceptual elements that are central to tissue engineering today: the use of polymeric materials in biological applications, the use of live cells to restore lost physiologic function, the vision of controlling tissue morphogenesis through the use of signaling molecules, and the powerful motivation of the shortage of transplantable organs.

Further, the seminal development in awareness of tissue engineering in the scientific community as a distinct domain of research is also attributable to Langer and Vacanti. The October 1987 Panel Meeting at NSF and the subsequent Granlibakken workshop in 1988 and Keystone workshops in 1990 and 1992, while significant, appear to have "spoken" to a much narrower audience; among our group of interviewees, few who were not directly involved in one or the other of these events report any awareness of their direct impact, although, again, the field definition drafted at the October meeting is acknowledged.

NSF's Directorate for Engineering also appears to have played a large role in incorporating the biomechanics community into the emerging field. The events of 1987-88 engaged pioneers of an engineering approach to orthopedics and hemodynamics, such as Richard Skalak, Van Mow, and Robert Nerem, in the development of a shared concept for the field just as the emergence of the scaffolds-and-cell-seeding approach, occurring in parallel with and independently of the Foundation's activities, was providing a renewed and powerful impetus for efforts to construct cell-based therapeutic solutions.

One might also note the delayed efforts of the Foundation to engage biologists in the endeavor. All of the participants in the Foundation's 1987 discussions were well-aware of the importance of fundamental research in biology as part of the mix required to make tissue engineering a success. As an agency concerned with the full range of fundamental research and not directly with clinical applications, NSF might have been the natural locus for an effort to mobilize basic biological scientists as well. However, as an NSF initiative, tissue engineering arose from the engineering directorate, and judged by data available from the NSF Fast Lane grants database, remained largely contained within the engineering directorate.

9

Research on Bamboo Micropropagation

For bamboo different propagation techniq ues are available, such as seed propagation, clump division, rhizome and culm cuttings. But these methods suffer from serious drawbacks for large or mass scale propagation. For mass scale propagation (> 500 000 plants per year) classical techniques are largely insufficient and inefficient, and tissue culture is the only viable method. Indeed, the order of magnitude of the demand for bamboo planting materials indicates that micropropagation will inevitably be necessary for mass scale propagation.

By now a large number of papers on micropropagation (the use of tissue culture for propagation only) of bamboos have been published, original papers as well as reviews and some in which tissue culture is described in a more general aspect. Many researches have focussed on somatic embryogenesis of seedlings of tropical bamboos. An inventory shows that at least 21 labs in South-east Asia were involved in bamboo tissue culture, mainly in India. Major research focused also on the clonal propagation of elite genotypes, either juvenile or adult. The number of papers about this subject however is much less, and this is solely due to lack of success. Indeed, technically the propagation of adult plants via axillary branching is much more difficult than with seedlings of tropical bamboos.

For tissue culture of bamboo the use of starting material (seeds or adult plants) and the choice of the propagation method are crucial. The two major advantages of using seedlings are that seedlings establish a new generation, and that the techno logy is easier. But the disadvantages are considerable:

- Insufficient or no knowledge of genetic background,
- Restricted availability of seeds for most species and rapid loss of germination capacity,
- Comparison of in vitro to in vivo performance has not been thoroughly evaluated.

In addition there is a huge variability in responsiveness in tissue culture.

When using adult bamboos main problems are:

- Endogenous contamination,

- Hyperhydricity and instability of multiplication rates,
- Many problems with rooting also in bamboos that root readily in nature.

Rooting percentages for adult bamboos ranged from very low percentages of 10 per cent for Bambusa vulgaris to 73 per cent for adult Dendrocalamus longispathus. A rooting percentage of 77 per cent was obtained for adult Dendrocalamus giganteus in 3 or 4 weeks. But, while 77 per cent of success is good on 500 plants in a laboratory experiment it still represents a loss of 33000 when we transplant 100 000 plants. Low rooting frequencies are the major bottleneck to developing commercially viable protocols. The combination of photomixotrophic in vitro multiplication and photoautotrophic in vitro rooting stages resulted in improved transplanting success. Improvement of rooting percentages and transplanting has been achieved in various commercial laboratories.

FLOWERING OF BAMBOO IN TISSUE CULTURE

Because of the peculiar flo wering habits in bamboo it has been almost impossible to breed for superior traits in woody bamboos. The first reports on tissue culture flowering of bamboos caused great excitement. It opened up the possibility of controlled flowering that can be used for breeding of bamboo. Since then in vitro flowering has been observed in many types of bamboos, both in seedlings and mature bamboos. More than 10 years after the first report on in vitro flowering in bamboo some positive results have been obtained, but practical and commercially exploitable results have not been reported yet. The most promising potential is for hybridisation under controlled conditions. But many hurdles still need to be taken before the methods really become applicable at agricultural scales.

Recently a new research project was initiated in our group to improve the induction of flowering and the quality of flowering parts, in order to be able to hybridise bamboos in the future. While this project aims mainly at developing techniques, it also includes various approaches on a more fundamental level. The use of pseudospikelets is central in this research. Pseudospikelets occur only in bamboos, not in other grasses. In grasses a spikelet bears two empty glumes, but the glumes of pseudospikelets in bamboos subtend dormant buds that can develop into new pseudospikelets. The proximal parts of the pseudospikelets do not develop into flowers but in tissue culture conditions keep on multiplying indefinitely, allowing to establish monocultures of pseudospikelets.

Flowers are formed in the distal parts of individual pseudospikelets and do not develop to anthesis under normal tissue culture conditions. They can however, be forced to develop fully under more adverse conditions. Cultural conditions can control flowering in vitro to a large extent. Factors used for the induction of pseudospikelets are mainly cytokinins. However, cytokinins are involved in various flowering processes and more fundamental research is urgently needed. Unfortunately, despite the huge potential of tissue culture flowering of bamboo, to our knowledge only one other laboratory is actively engaged in this field of research.

Despite the presence of flowers in the axils of pseudospikelets, dormant buds can develop into vegetative shoots. Pseudospikelets can also be regarded as highly contracted axes, in which leaves are reduced to glumes. Instead of propagules and plantlets of 2 to 5 cm in tissue culture, pseudospikelets have lengths between 0.25 and 1.5 cm. This is a miniaturisation of the plants and allows a good opportunity to use these pseudospikelets for propagation of bamboos.

With these smaller clumps, operators can cut and transfer much easier, while the mode of propagation is still via axillary branching, thus the method with lesser risk for somaclonal variation. Since leaf area is much reduced, quality problems in the propagation stage are much less important. This allows for a considerable cost reduction in micropropagation, since operator costs represent 75- 80 per cent of cost price of tissue culture plants. In tissue culture systems for propagation flowering of bamboo represents considerable risks since many bamboos flower monocarpically in nature and die completely after flowering. It is very important to be able to control flowering and reversion of flowering of bamboo. Therefore the influence of tissue culture flowering on vegetative growth post vitro has been studied extensively in three independent experiments (unpublished results).

- Plants of Bambusa tuldoides regenerated from pseudospikelets grow as vegetative plants after hardening;
- In large scale vegetative propagation only two flowering plants of Phyllostachys have been observed after transplanting about 500000 plants,
- Tests with Sasa palmata have shown that even when flowering plants are transplanted in the greenhouse, these revert to the vegetative state within a few weeks.

These tests have shown that tissue culture technology does not lead to monocarpic flowering of saleable plants. Using pseudospikelets as propagules we have now a system which can be used for propagation and for the induction of flowering and flowering structures. The same propagules can also be used for long term storage and for genetic transformation.

COMMERCIAL MICROPROPAGATION OF BAMBOO

Micropropagation via tissue culture thus attracted a lot of attention but the translation and transformation of these expectations into commercially viable propagation systems has been beset with a number of problems that were either technological, or related to marketing. At present only a very limited number of laboratories specialise in the production of bamboos at a commercial scale. Oprins Plant, Belgium, currently has over 60 genotypes in culture initiated from adult genotypes of which about 30 on a commercial scale.

Species and varieties of Arundinaria, Chimonobambusa, Fargesia, Phyllostachys, Pleioblastus, Sasa, Sasaella, Semiarundinaria, Shibataea and Yushania (temperate bamboos) and Bambusa, Dendrocalamus, Dinochloa, and Thyrsostachys (tropicals) are produced commercially. Rooting percentages of most species are between 85 and 100 per cent, depending on species and season of transplanting (unpublished results). These bamboos are produced either as ornamental, or for tropical forestry. Other laboratories focusing on bamboo tissue culture include West Wind Technology, USA, and Bamboo World, Australia.

Several other laboratories such as Piccoplant, Germany, and Microflor, Belgium, produce a more limited number of bamboo species as part of their tissue culture operation. Other laboratories initially involved in tissue culture of bamboos however, no longer produce bamboo commercially, for example TERI, India and Thai Orchids Lab, Thailand. The main problem is that micropropagation of bamboos also has to deal with the development of new markets and market opportunities. Logistics and supply chains such as in ornamental horticulture where tissue culture labs are separate entities seem not to work. Instead the laboratories have to develop their own markets, marketing and sales for bamboo plants. This strategy of forward integration needs to go beyond the classical scheme of tissue culture. The tissue culture phase terminates at Stage III, but in regard to marketing of plants one can also distinguish subsequent stages:

- Stage IV, the transplantation stage with the end product being rooted plantlets in trays,
- Stage V, the production of liners, either for production of saleable plants or for use as micromotherplant,
- Stage VI, the production of saleable plants.

This distinction is important if the complete chain of production is integrated in a single company, since this determines the added values. In our case the laboratory produces ornamental plants exclusively for Oprins Plant's own nurseries. A balanced production and logistic chain, where bamboos are produced in Belgium and grown in the nurseries in Spain and France, has allowed to produce high quality ornamental bamboos year round at prices which are 40-60 per cent below the current market prices for bamboo. These plants are currently being sold under the brand name BambooSelectTM, which is an additional tool in marketing. In an agricultural perspective it is now possible to produce the same genotypes for large scale plantations in Europe at prices comparable to other energy crops.

MONITORING GENETIC STABILITY IN BAMBOO TISSUE CULTURE

AFLP Markers in Quality Control of Micropropagation

Somatic mutations in bamboo are highly valued in horticulture. In various bamboos somatic mutations are known that alter the stem colour or stem shape, such as Bambusa ventricosa with inflated internodes. But not all are stable in tissue culture.

The bulbous internodes of Bambusa ventricosa were lost after tissue culture and variegation of stems and leaves disappears very often in tissue culture. Tissue culture has been associated with molecular aberrations in general (including also chromosomal defects), and phenotypic defects may show up only some years later. Therefore it is important to devise methods of quality control at the tissue culture stages to minimise the potential danger of such defects, for exa mple when plants of Stage VI are sold in large quantities.

Generally it is very difficult to find precise defects, especially when somatic mutations are possibly linked to transposon activity. In a first approach AFLPTM markers are used to assess the stability and clonal fidelity of tissue cultured bamboos. In one experiment Phyllostachys species and cultivars were collected:

- From the garden of Dr. Jacques Van Dooren, Kumtich, reference collection of bamboo for Belgium
- From the tissue culture lab at Oprins Plant
- from plants transplanted in the greenhouse.

Using this set up, it should be possible to detect if genotypes in the collection and in propagation have the correct names, and whether the plants cultured are subject to considerable clonal variation.

The most bamboos in tissue culture and transplanted in the greenhouse group together with the collection plants. The exception is Phyllostachys vivax from the laboratory that turns out to belong to the nigra group, while the collection plant of P. vivax does not group at all with the nigra's.

This is a clear case of mistaken identity, due to mislabelling of the motherplants obtained from China. If one observes the dendrograms it is seen tha t all P. aurea cultivars, all P. nigra cultivars, and all P. aureosulcata cultivars group together. They are, according to this test, 100 per cent identical. However, they differ in stem colour, leaf colour, habit etc, and these differences can be observed readily in the field. This also means that the method used is not sensitive enough to detect those differences. But it clearly shows that AFLP is a useful method when used as part of quality control in the tissue culture process, to assess correct clonal identity. This can then be used towards customers. AFLP markers are now used routinely besides other measures to ensure clonal fidelity. These other measures include the use of axillary branching as propagation method and annual initiation of new cultures, so that the maximum number of subcultures is 12 to 14.

MOLECULAR BASIS OF SOMATIC MUTATIONS IN BAMBOO

Based on phenotype and mode of action somatic mutations involved in stem colour of bamboo are possibly caused by transposon activity. The study of transposon activity and their possible relation to certain mutations is very difficult. In bamboo no breeding systems are available to detect transposon activity through classical breeding. Molecular approaches may provide some shortcuts, but to link the presence and activity of transposons to certain phenotypic markers is again a difficult task.

A first approach to identify the molecular basis of these somatic mutations was to use primers based on sequence information from the 4.5 kb Ac9 transposon from maize. Several primer combinations were used and several copies with considerable homology to the original Ac9 transposons were present, which eventually turned out to be a useful and relatively cheap method to identify bamboos, at the species level.

Primers designed to detect Activator-like transposons in Petunia could also be used to detect fragments in bamboo. The same fragment was detected in the three bamboos tested, namely Bambusa vulgaris, Sasa veitchii and Phyllostachys edulis. The fragment of Bambusa vulgaris (hATbv1) was sequenced and shown to be homologous to members of the hAT superfamily of transposons, to which also Ac (corn), Tam (Antirrhinum majus) and hobo (Drosophila) belong. The homology was as high as 60 per cent in some regions, which conforms also to earlier findings of Ac-like sequences in Bambusa multiplex. It is very likely that many other transposons are present in the bamboo genome as well.

A next step is to use these AFLP-primers in combination with methylation sensitive AFLP, in which methylation sensitive restriction enzymes are used to cleave sites within the bamboo transposon. The relation between methylation patterns and transposon activities is well established and the activation of transposons during tissue culture is also a well documented phenomenon.

The use of genotypes that have been cultured for six, eighteen, twenty-four and thirty-six months should allow to assess if long term tissue culture has any influence on methylation in general and on methylation patterns of transposons. This research is currently ongoing.

10

Genetic Engineering in Plant Breeding

AIM OF PLANT BREEDING

The constant aim of plant breeding is to improve the quality, diversity and performance of agricultural and horticultural crops. The overriding objective is to develop plants better adapted to human needs. The evolution of plant breeding is a classic example of how improved biological understanding has been adapted to provide more effective methods of meeting the demands of a changing world. Plant breeding has existed in its most primitive form since the first farmers saved the seeds of their best plants from one season to the next more than 10,000 years ago. Over the centuries, this selection process has gradually become more scientific, bringing major improvements in the yield, quality and diversity of crops grown worldwide.

In the 19th century Gregor Mendel established the basic principles of plant genetics. Gregor Mendel in the 19th century laid the foundations for traditional plant breeding, in which selected parents are cross-pollinated to combine certain desired characteristics, such as high yield and resistance to pests and disease. He discovered that units of material, which are transferred from one generation to the next, determine inherited traits.

The plant breeder's aim is to reassemble these units of inheritance, known as genes, to produce crops with improved characteristics. In practice, this is a complex and time-consuming process. Each plant contains many thousands of genes, and the plant breeder is seeking to combine a range of desirable traits in one plant to produce a successful variety. Conventional breeding involves crossing selected parent plants, chosen because they have desirable characteristics such as high yield or disease resistance. The breeder's skill lies in selecting the best plants from the many and varied offspring. These are grown on and tested in subsequent years. Typically this involves examining thousands of individual plants for different characteristics ranging from agronomic performance to end-use quality. Developing a new variety can take up to 15 years for wheat, 18 years for potatoes, even longer for some crops.

The scope of conventional plant breeding has increased with improvements in technology. In the laboratory, chemical and mechanical techniques are used to speed up the selection process and remove natural barriers to cross-fertilization, for example between different crop species.

All forms of plant breeding involve the improvement of crops by selection. Since man began farming he has selected seeds from the best plants for the next generation. Over the centuries, this selection process has become more scientific, bringing major improvements in the yield, quality and diversity of agricultural and horticultural crops. Genes-units of hereditary material that are transferred from one generation to the next, determine these characteristics. Since the plant contains many thousands of genes, and the breeder is seeking to combine a range of traits in one plant, developing a successful variety can be an extremely lengthy process-up to 12 years in the case of cereals.

Using biological knowledge derived from cross-pollination, plant breeders have developed physical and chemically assisted ways of enhancing the speed, accuracy and scope of the selection process. Breeding systems involving grafting and hybridization, for example, have extended breeders' capabilities at the whole crop level, while more recent cell-based techniques enable breeders to operate at the level of individual cells and their chromosomes

A major objective in modern plant breeding is the making of crop varieties with the highest possible yield potential. Yield potential is defined as "the yield of a crop when growth is not limited by water or nutrients, pests, diseases, or weeds". For farmers, whose crops are indeed limited by such constraints, this yield concept may not be seen as the most relevant objective. Relieving crops of all sorts of environmental stress, leaves light and temperature, together with varietal characteristics, as the only determinants of yield. That achieved, the same high yielding variety can be used all over a climatic zone, and target area for the variety is enormous. If stress can not be eliminated, however, adaptation to a usually site-specific environment becomes necessary. In that case, the target area for each variety will be small.

Do these different perspectives warrant different approaches to plant breeding? Conventional wisdom says no. In experiments where varieties are tested at various levels of inputs, the same high yielding variety usually comes out as top yielder at all levels. Therefore, plant breeders often claim that their varieties not only perform excellently under high-input conditions, but would also be better than traditional varieties in a low-input environment. However, when these trials are taken outside of experimental farms and the varieties are tested under local or farm conditions, researchers discover what is called 'crossover in performance'. At a certain level of stress there is a crossover point beyond which local varieties or landraces perform better than the high yielding varieties Farmers who experience such situations are not a tiny minority.

To some degree, most if not all farmers have to cope with local stress conditions. The stress-free environment is hard to achieve, and even harder to sustain. In many favourable areas, farmers abandon the high-input technology for economic reasons or because of ecological problems, thus increasing the need for locally-adapted germplasm. But modern plant breeding, whether public or private, can not supply adapted germplasm everywhere. Only a system of local seed selection can ensure that. And that means devolution of plant breeding.

Can such decentralized breeding be compatible with development needs in a changing world and meet economic aspirations in a poor society? If the answer to these questions is going to be 'yes', the decentralized breeding must be able to take advantage of the power of science as well as of the capacities of local communities. Three aspects should be considered and made mutually compatible: breeding technology, participatory research methods, and organization at community level.

HISTORY

Plant breeding is the purposeful manipulation of plant species in order to create desired genotypes and phenotypes for specific purposes. This manipulation involves either controlled pollination, genetic engineering, or both, followed by artificial selection of progeny. *Plant breeding* often, but not always, leads to plant domestication.

Plant breeding has been practiced for thousands of years, since near the beginning of human civilization. It is now practiced worldwide by government institutions and commercial enterprises. International development agencies believe that breeding new crops is important for ensuring food security and developing practices of sustainable agriculture through the development of crops suitable for their environment.

Domestication

Domestication of plants is an artificial selection process conducted by humans to produce plants that have fewer undesireable traits of wild plants, and which renders them dependent on artificial (usually enhanced) environments for their continued existence. The practice is estimated to date back 9,000-11,000 years. Many crops in present day cultivation are the result of domestication in ancient times, about 5,000 years ago in the Old World and 3,000 years ago in the New World. In the Neolithic period, domestication took a minimum of 1,000 years and a maximum of 7,000 years. Today, all of our principal food crops come from domesticated varieties.

A cultivated crop species that has evolved from wild populations due to selective pressures from traditional farmers is called a landrace. Landraces, which can be the result of natural forces or domestication, are plants (or animals) that are ideally suited to a particular region or environment. An example are the landraces of rice, *Oryza sativa* subspecies *indica,* which was developed in South Asia, and *Oryza sativa* subspecies *japonica,* which was developed in China.

Classical plant breeding uses deliberate interbreeding (crossing) of closely or distantly related species to produce new crops with desirable properties. Plants are crossed to introduce traits/genes from one species into a new genetic background. For example, a mildew resistant pea may be crossed with a high-yielding but susceptible pea, the goal of the cross being to introduce mildew resistance without losing the high-yield characteristics. Progeny from the cross would then be crossed with the high-yielding parent to ensure that the progeny were most like the high-yielding parent, (backcrossing), the progeny from that cross would be tested for yield and mildew resistance and high-yielding resistant plants would be further developed. Plants may also be crossed with themselves to produce inbred varieties for breeding.

Classical breeding relies on homologous recombination of two genomes to generate genetic diversity. The Classical plant breeder may also makes use of a number of *in vitro* techniques such as protoplas fusion, embryo rescue or mutagenisis to generate diversity and produce plants that would not exist in nature.

Traits that breeders' have tried to incorporate into crop plants in the last 100 years include:

- Increased quality and yield of the crop
- Increased tolerance of environmental pressures (salinity, extreme temperature, drought)
- Resistance to viruses, fungi and bacteria
- Increased tolerance to insect pests
- Increased tolerance of herbicides

Before World War II

Intraspecific hybridization within a plant species was demonstrated by Charles Darwin and Gregor Mendel, and was further developed by geneticists and plant breeders. In the early 20th century, plant breeders realised that Mendel's findings on the non-random nature of inheritance could be applied to seedling populations produced through deliberate pollinations to predict the frequencies of different types.

In 1908, George Harrison Shull described heterosis, also known as hybrid vigour. Heterosis describes the tendency of the progeny of a specific cross to outperform both parents. The detection of the usefulness of heterosis for plant breeding has lead to the development of inbred lines that reveal a heterotic yield advantage when they are crossed. Maize was the first species where heterosis was widely used to produce hybrids. By the 1920s, statistical methods were developed to analyse gene action and distinguish heritable variation from variation caused by environment. In 1933, another important breeding technique, cytoplasmic male sterility (CMS), developed in maize, was described by Marcus Morton Rhoades. CMS is a maternally inherited trait that makes the plant produce sterile pollen, enabling the production of hybrids and removing the need for detasseling maize plants. These early breeding techniques resulted in large yield increase in the United States in the early 20th century. Similar yield increases were not produced elsewhere until after World War II, the Green Revolution increased crop production in the developing world in the 1960s.

After World War II

Following World War II a number of techniques were developed that allowed plant breeders to hybridize distantly related species, and artificially induce genetic diversity. When distantly related species are crossed, plant breeders make use of a number of plant tissue culture techniques to produce progeny from other wise fruitless mating. Interspecific and intergeneric hybrids are produced from a cross of related species or genera that do not normally sexually reproduce with each other. These crosses are referred to as *Wide crosses*. The cereal triticale is a wheat and rye hybrid. The first generation created from the cross was sterile, so the cell division inhibitor colchicine was used to double the number of chromosomes in the cell. Cells with an uneven number of chromosomes are sterile.

Failure to produce a hybrid may be due to pre- or post-fertilization incompatibility. If fertilization is possible between two species or genera, the hybrid embryo may abort before maturation. If this does occur the embryo resulting from an interspecific or intergeneric cross can sometimes be rescued and cultured to produce a whole plant. Such a method is referred to as *Embryo Rescue*. This technique has been used to produce new rice for Africa, an interspecific cross of Asian rice *(Oryza sativa)* and African rice *(Oryza glaberrima)*. Hybrids may also be produced by a technique called protoplast fusion. In this case protoplasts are fused, usually in an electric field. Viable recombinants can be regenerated in culture.

Chemical mutagens like EMS and DMSO, radiation and transposons are used to generate mutants with desirable traits to be bred with other cultivars. Classical plant breeders also generate genetic diversity within a species by exploiting a process called somaclonal variation, which occurs in plants produced from tissue culture, particularly plants derived from callus. Induced polyploidy, and the addition or removal of chromosomes using a technique called chromosome engineering may also be used. When a desirable trait has been bred into a species, a number of crosses to the favoured parent are made to make the new plant as similar as the parent as possible. Returning to the example of the mildew resistant pea being crossed with a high-yielding but susceptible pea, to make the mildew resistant progeny of the cross most like the high-yielding parent, the progeny will be crossed back to that parent for several generations. This process removes most of the genetic contribution of the mildew resistant parent. Classical breeding is therefore a cyclical process. It should be noted that with classical breeding techniques, the breeder does not know exactly what genes have been introduced to the new cultivars.

Some scientists therefore argue that plants produced by classical breeding methods should undergo the same safety testing regime as genetically modified plants. There have been instances where plants bred using classical techniques have been unsuitable for human consumption, for example the poison solanine was accidentally re-introduced into varieties of potato though plant breeding. Exploiting heterogeneity and crop evolution in farmers' fields are outside the scope of most plant breeding research. One exceptional experiment, however, has shed some scientific light on the issue.

It was started at the University of California (UC) in 1928. Composite cross populations of barley were produced, some of which were extremely diverse in origin of sources. These populations were exposed to continuous natural selection in current modern farming environments and became the subject of studies during the career span of several generations of UC professors. It appears that after low yields in initial years, the composite cross populations gradually improved in performance and eventually became quite good yielders, with excellent yield stability and disease resistance. These results inspired Suneson to propose an evolutionary plant breeding method. After assessment of later generations of the same material, Soliman and Allard concluded that such evolutionary breeding "is unwarranted" if yield potential is the major goal.

However, if disease resistance and yield stability are two major objectives, "the composite cross approach is an efficient method". This amounts to saying that a major part of world agriculture, many high-input systems included, could be well served by this approach. In short, this means constructing a body of broadly diversified germplasm and exposing it to natural selection in areas of contemplated use. For those who are familiar with traditional farmers' breeding, this may sound like reinventing the wheel. In fact it is an improvement of the old wheel of plant breeding.

The first step, constructing a body of of broadly diversified germplasm, is not all that straightforward. Science has access to world collections and information sources that are unavailable to farmers. A research institute can chose relevant germplasm and make composite cross populations with an evolutionary potential, most probably far beyond that of locally available varieties.

The immediate outcome, the early generation composite cross population, will be unadapted everywhere and is likely to yield poorly. With time, however, recombinations and natural sorting will improve the adaptation, and, according to the Californian experience, narrow the gap with commercial varieties. The long term outcome could be populations that outperform commercial varieties in disease resistance and yield stability and that may be used as a source of artificial selection for high yield. The disease resistance appears to have evolved through the building up of polygenic complexes. Therefore, it provides a durable resistance as opposed to the monogenic and, therefore, mostly non-durable resistance usually bred into commercial pure line varieties. The stability, at least to some degree, depends on the buffering effect of crop heterogeneity. The Californian experiment shows that when the population is propagated in isolation for a very long time, diversity will start declining, resulting eventually in reduced stability.

These experimental findings into the context of current development needs, a few conclusions can be drawn. We need breeding populations with a very high evolutionary potential, and these populations must be exposed to the stress conditions of, or similar to, current farm environments. Furthermore, a certain level of diversity within populations must be maintained in order to sustain evolutionary potential and yield stability.

CONVENTIONAL BREEDING

Conventional plant breeding involves crossing carefully chosen parent plants, then selecting the best plants from the resulting offspring to be grown on for further selection. For cereals, hundreds of individual crosses are carried out by hand to create seed for the first filial (or F1) generation. The resulting F1 plants are uniform, but in the following generation several hundred thousand different plants are produced. Because of the way genes work, the new combinations produced from each cross are not revealed until the second (F2) generation.

It is this enormous diversity of new gene combinations that may hold the key to a successful new variety. The plant breeder's task is to select the plants most likely to meet his breeding objectives. Seed from the best of these F2 plants is grown on in small rows or plots and the best plants again selected– this process is repeated year after year until only the very best plants remain. As promising new lines emerge, tests arc conducted on each plot to assess factors such as yield, disease resistance and endues quality. Once the best lines are purified to ensure that every plant has the same characteristics, the process of multiplying seed begins. These 'inbred' lines are then ready for entry into official trials, some six to ten years after the initial cross.

HYBRID BREEDING

For some crop species, the seed supplied to growers is that produced from the first cross between selected parents. The resulting varieties, known as F1 hybrids, offer potential advantages in crop performance. Hybrid breeding is widely used to produce varieties of field vegetables, maize and oilseed rape. F1 hybrids are unique in expressing 'hybrid vigour' in the growing crops for a single year. This may result in higher yields, greater uniformity, or improvements in quality. Unlike inbred lines, F1 hybrids do not breed true year after year and their performance gains are not maintained in subsequent generations.

IMPROVED BREEDING

With increasing knowledge and improved technology, breeders have developed ways to enhance the speed, accuracy and scope of the breeding process.

There are several ways to reduce the lengthy interval between the first cross of selected parents and establishing true breeding lines of promising new varieties. For example, maintaining parallel selection programmes in northern and southern hemispheres allows two generations to be produced each year. Single seed descent produces very small plants under restricted growth conditions. Large numbers of these plants are cultivated in artificial growth rooms, with two or more generations produced in a year. In potatoes, mini-tuber breeding speeds up the slow multiplication process by producing miniature plants under greenhouse conditions.

More recent laboratory techniques enable breeders to operate at the level of individual cells and their chromosomes. In certain crop species, such as potatoes and oilseed rape, it is possible to produce new varieties in the laboratory through

Protoplast Fusion

Individual plant cells with their outer walls removed are fused, and the fused cells then induced to divide and grow in a culture medium. Whole plants are eventually regenerated containing new combinations of genes from the two parents.

Embryo Rescue and Assisted Pollination

Allow breeders to expand the range of available characters by making crosses between species, which would not produce viable offspring outside the laboratory.

Double Haploid Breeding

Enables breeders to produce genetically uniform lines within one generation. This effectively by-passes the lengthy process of self pollination and selection normally required to produce true breeding plants. Latest developments in genetic science have greatly improved our understanding of how plants behave, offering additional ways to enhance the breeding process.

Genomics

By mapping the genetic makeup, or genome, of crop species, scientists can identify the exact position and function of individual genes. Genome mapping has revealed striking similarities in the genomes of different crop species, such as rice, wheat, barley and rye. This information is already helping to broaden the scope and precision of current breeding programmes.

Marker Assisted Breeding

Allows breeders to determine whether desired traits are present in a new variety at an early stage in the breeding programme.

GENETIC MODIFICATION

Genetic modification of plants is achieved by adding a specific gene or genes to a plant, or by knocking out a gene with RNAi, to produce a desirable phenotype. The resulting plants are often referred to as transgenic plants. Genetic modification can produce a plant with the desired trait or traits faster than classical breeding because the majority of the plant's genome is not altered. To genetically modify a plant, a genetic construct must be designed so that the gene to be added or knocked-out will be expressed by the plant. To do this, a promoter to drive transcription and a termination sequence to stop transcription of the new gene, and the gene of genes of interest must be introduced to the plant. A marker for the selection of transformed plants is also included. In the laboratory, antibiotic resistance is a commonly used marker: plants that have been successfully transformed will grow on media containing antibiotics; plants that have not been transformed will die. In some instances markers for selection are removed by backcrossing with the parent plant prior to commercial release.

The construct can be inserted in the plant genome by genetic recombination using the bacteria Agrobacterium tumefaciens or *A. rhizogenes*, or by direct methods like the gene gun or microinjection. Using plant viruses to insert genetic constructs into plants is also a possibility, but the technique is limited by the host range of the virus. For example, Cauliflower mosaic virus (CaMV) only infects cauliflower and related species. Another limitation of viral vectors is that the virus is not usually passed on the progeny, so every plant has to be inoculated.

The majority of commercially released transgenic plants, are currently limited to plants that have introduced resistance to insect pests and herbicides. Insect resistance is achieved through incorporation of a gene from Bacillus thuringiensis (Bt) that encodes a protein that is toxic to some insects. For example, if cotton pest the cotton bollworm feeds on Bt cotton it will ingest the toxin and die. Herbicides usually work by binding to certain plant enzymes and inhibiting their action. The enzymes that the herbicide inhibits are known as the herbicides *target site*. Herbicide resistance can be engineered into crops by expressing a version of *target site* protein that is not inhibited by the herbicide.

This is the method used to produce glyphosate resistant crop plants.Genetic modification of plants that can produce pharmaceuticals (and industrial chemicals), sometimes called *pharmacrops,* is a rather radical new area of plant breeding

Proteomics

Allows breeders to understand how genes behave in different parts of the plant and under different growing conditions.

Maintaining Genetic Diversity

Maintaining biodiversity is central to the process of crop improvement. It is in every breeder's interest to ensure that the gene pool from which new traits are selected remains as extensive as possible. Plant breeders created the first gene banks in the 1930s to conserve the valuable genetic diversity within past and present varieties, as well as landraces and wild relatives of cultivated crop species. Plant breeding is integral to ongoing initiatives to identify, classify and conserve existing biodiversity.

Testing Plant Varieties

Before any new crop variety can be placed on the market, it must undergo statutory testing under a process known as National Listing. Successful varieties are placed on a National List or register of varieties approved for marketing. National Listing rules are determined on a European basis, and apply to all the major agricultural and vegetable crop species.

Official trials are conducted, in most cases for a minimum of two years, to test each new variety for a range of characteristics which together determine its uniqueness, its genetic uniformity, and its value to growers and the rest of the food chain. National Listing is extremely rigorous–the majority of varieties entered do not complete the process. In winter wheat, for example, only a quarter of varieties entered for National Listing during the 1990s were finally approved.

National Listing–DUS

All varieties entered for National Listing are assessed for Distinctness, Uniformity and Stability (DUS). In the case of cereals, some 30 individual characteristics of the plant are inspected to verify that it is distinct, ie clearly distinguishable from other varieties, that its characteristics are uniform from one plant to another, and that the variety is stable in that it breeds true to type from one generation to the next.

National Listing–VCU

For agricultural crops, National Listing also involves trials to establish a variety's Value for Cultivation and Use (VCU). This provides an assurance to growers that only varieties with improved performance or end-use quality can be officially approved for sale.

Recommended and Descriptive Lists

Once a new variety has been added to the National List, it is cleared for marketing. However, its success in the market place is by no means guaranteed. Further independent trials are conducted each year to compare the performance and quality of the best varieties. These trials provide the basis for detailed information and advice to growers and their customers.

As shown in the following illustration for winter wheat, the process of statutory and commercial evaluation can take up to five years. Only a handful of varieties clear all these hurdles, which are in addition to the many years spent testing and selecting in the breeder's own trials programme.

Genetically Modified Varieties

No GM crops can be marketed until they have been assessed and approved in terms of human health, food safety and the environment. The cross-border movement of genetically modified organisms (GMOs) is regulated at a global level under the internationally agreed Biosafety Protocol.

Variety Maintenance and Identity Preservation

As well as developing new varieties, plant breeders maintain the genetic purity of existing lines and pre-commercial seed supplies year by year. This process is costly and time-consuming, but essential to maintain the quality and performance of each variety.

For cereals, variety maintenance begins after a few years of selection trials, when the promise of a variety is just emerging. At that stage, all that exists of what may become a widely grown variety is a single row containing around 100 plants. The breeder then bulks up supplies of the purified lines of breeder's seed into prebasic and then basic seed. Each year specialist seed growers are used to grow basic seed for the first generation of certified seed–or C1 seed. After one more year this becomes C2 seed, the main source of certified seed used by farmers.

The plant breeder continuously maintains breeder's seed for the process of multiplication through pre-basic, basic, C1 and C2 seed to ensure the variety's performance and quality year after year. Greater emphasis is now being placed on preserving the identity of individual varieties after harvest, both to conserve quality characteristics and to meet consumer demands for assurances about the integrity and traceability of their food.

Seed Certification

Seed of an approved variety can only be marketed if it meets strict quality criteria Seed quality standards are laid down in UK and EU law, and policed by agencies appointed by Government. The UK's official seed certification system offers an independent assurance of quality to growers. Minimum standards apply for varietal identity, purity and germination capacity. In addition, strict limits apply to seed-borne diseases, and the presence of physical impurities such as weed seeds.

Around 9% of the UK arable area is used to multiply the pure lines of seed from the plant breeder into certified seed. Several thousand individual crops are involved; each grown under specific management regimes to ensure the purity and integrity of the resulting seed is maintained. To gain certification, every seed crop must undergo crop inspection and seed testing. Seed certification underpins the health and purity status of the major arable crops in Britain. It offers an independent benchmark of quality on which buyers of seed and their customers depend.

Farm-saved Seed

For certain crop species–particularly small-grain cereals–growers can opt to save their own seed for sowing the following year provided care is taken to ensure that the crop remains healthy and free from impurities, and that the resulting seed is carefully conditioned and cleaned. Without independent testing for germination and freedom from seed borne diseases, however, farm-saved seed can harbors risks to growers that may not become apparent until well into the growing season. Many growers choose certified seed for the peace of mind that quality and performance is independently assured.

Plant Breeding in the Future

Success in any line of business depends on being able to anticipate and respond to changing market requirements. Such is the long-term nature of developing a new crop variety that plant breeders are in the unenviable position of forecasting their customers' demands at least five, and probably nearer ten years, in advance.

Plant breeding objectives will continue to be shaped by changes in agricultural and environmental policy. There is no doubt that while improvements in crop yield and endues quality remain of paramount importance, the future for British agriculture lies in crop production systems which deliver both economic and environmental benefits. Increasingly, plant breeders are targeting varieties suitable for low input and organic regimes. It is already estimated that genetic improvements in disease resistance alone save British farmers more than £100 million a year in reduced agrochemical use and improved yield. A range of other factors also determines a variety's suitability for low input, integrated farming systems. They include the crop growth cycle, straw strength and susceptibility to weed competition, as well as the variety's ability to thrive in a range of soil types, climatic conditions and rotational patterns.

Novel Crops

Breeders continue to adapt novel crops for UK growing conditions. New varieties of sunflowers and soya beans are already creating market opportunities in the food sector. Post-BSE, attention has focused on using more home-grown fodder in animal feed. For plant breeders, this means enhancing energy and protein levels in traditional forage crops, and developing the potential of new crop species, such as lupins.

Non-food Crops

The use of crop plants for industrial, nonfood applications is an area of increasing interest. Developments in plant breeding could underpin a transition from many crude, industrial processes towards renewable, field-based production of specialist chemicals, textiles and biofuels Quick-growing biomass crops such as willow coppice and miscanthus are already fuelling UK power stations, and there is huge potential to develop oilseed, starch and fibre crops for a wide range of product including lubricants, pharmaceuticals, dyes, plastics, flooring, paper and clothing.

New Traits

Pioneering research by British scientists is opening up new possibilities for crop improvement. Examples from current research include the use of genetic modification to overcome grey mould infestation in strawberries–not possible through conventional breeding because no resistance genes exist in the world strawberry germplasm. Similar techniques offer ways to control the ripening process in fruit and vegetables, to extend storage life, and improve flavour and texture for consumers. It will also increase the seasonal availability of homegrown produce, so reducing the distance food is transported. Modern biotechnology can be used to improve human nutrition. Recent research by British scientists includes the development of vegetables containing increased levels of cancer-fighting nutrients, and apples, which really do keep the dentist at, bay by protecting against tooth decay.

SCOPE OF PLANT BREEDING

IMPROVING CROP PERFORMANCE

The dramatic gains in agricultural productivity seen in the second half of the last century are often linked uniquely to increased mechanisation and the widespread introduction of fertilizers and agrochemicals.This 'Green Revolution' would not have been possible without the huge genetic improvements made to the crops. Plant breeding has contributed around half of the threefold increase in UK wheat yields recorded from 1947 to 1992 and yields have continued to improve since then. Plant breeding can directly improve the performance of crops in different ways.

Productive Yield

Developing crop varieties, which convert more of their biomass into productive yield, is the single biggest contributor to improved crop output. The introduction of shorter-strawed cereals is a striking example of how this has been achieved, by transforming more of the crop's productive energy into valuable grain.

Physical Characteristics

Changing a crop's physical structure can also contribute to increased yields. For example, the development of semi-leafless varieties of field peas helped to boost intrinsic yield. It also stimulated the growth of stabilising tendrils, which significantly improved the crop's standing ability so reducing crop losses at harvest.

Disease Resistance

Genetic resistance to disease enables crops to realise their yield potential– it can also mean reduced use of agrochemicals. Plant breeding has significantly improved the genetic resistance of crops against the threat of viral and fungal infection. Key examples include resistance to blight in potatoes, rhizomania in sugar beet, and barley yellow mosaic virus in cereals. The challenge of breeding resistant varieties is constant because new strains of disease develop naturally.

Timing of Maturity

Plant breeding technology has brought major improvements in the uniformity with which crops ripen ready for harvest. This not only reduces potential crop losses at harvest (as in the case of pod shatter in oilseed rape), but has also improved growers' ability to mechanise harvesting operations. In the field vegetable sector, for example, the labour intensive (and unpleasant!) task of picking crops such as cauliflower, broccoli and Brussels sprouts has been transformed by the development of improved varieties.

Other Agronomic Factors

By improving crops' ability to cope with a range of other agronomic pressures, advances in plant breeding continue to underpin progress in agricultural productivity.

They include:

- Genetic resistance to pests, such as nematode resistance in potatoes
- Shorter crop life-cycle, to expand a crop's geographical growing areas
- Stress tolerance, such as frostresistance in field vegetables, to extend the seasonal availability of home-grown fresh produce.

Improving Crop Quality

Alongside progress in keeping Britain's farmers competitive through improved productivity, developments in plant breeding have also brought significant gains in crop quality. Gone are the days when Europe encouraged farmers to concentrate solely on maximising output, assured that their crops would find support under the Common Agricultural Policy.

Today, as production-based support and barriers to international trade continue to be reduced, Britain's growers must compete with the best in the world, on quality and cost of production. Consumer expectations of food are also more exacting. This is reflected in tighter quality specifications along the food chain. Plant breeders have responded with a continuous stream of new varieties, tailored to the needs of specific end-markets.

Cereals

Britain's historical dependence on North American imports of wheat for bread making has been reversed over the past 40 years thanks to improvements in baking technology and the development of high protein, hard milling varieties of wheat.

Until the mid-1960s, imported wheat typically accounted for up to 65% of our bread making requirements. Today as much as 90% of our bread is produced from homegrown varieties. In barley, improvements in malting quality have underpinned a fourfold increase in the volume of beer brewed per tonne, from around 2,000 liters in 1950 to nearly 8,000 liters today.

The quality of malting barley varieties grown in Britain is internationally recognized, with export markets from the near Continent to the Far East.

Oilseed Rape

Oilseed rape has become a major homegrown source of vegetable oil and animal feed within the past 30 years. The rapid expansion of this important break crop from 50,000 hectares in the mid-1970s to around 500,000 hectares today can be linked to two major breakthroughs in plant breeding. Varieties for food use were first developed in 1970. These contain reduced levels of erucic acid in the oil, making it more suitable for human consumption. This was followed in the late 1980s by 'doublelow' varieties, which also offered reduced levels of glucosinolates in the residual meal to improve quality for animal feed.

Potatoes

In the potato sector, plant breeders have responded to increasing consumer demands for quality and choice. Fresh produce counters now boast a bigger range than ever, from traditional main-crop varieties through to baby new, baking and salad potatoes. Breeding programmes have also targeted processing and catering markets, with specific varieties used to produce crisps, instant mash and chips.

Sugar Beet

Higher root yield and increased sugar content have combined to double sugar production per hectare in the past 50 years. The introduction of monogerm seed, allowing seeds to be planted individually rather than in clusters, ensures genetic improvements are fully realised in the field.

Pulses

As interest in home-grown sources of protein has increased, breeders have developed varieties of peas and beans for specialist markets–freezing and canning for human consumption, flaking for pet food, and tannin-free varieties for animal feed.

Creating a New Variety

The creation of each new variety is a complex, costly and skilled operation. It is also painstakingly slow–today's breeding programmes are already looking ten years ahead to the needs of farmers, consumers and the environment at the end of the next decade and beyond. Techniques vary between crop species, but the scientific principles of plant breeding remain unchanged from Mendel's first discovery that selected parent plants can be cross-pollinated to combine desired characteristics. Genes–units of hereditary material that are transferred from one generation to the next, determine plant characteristics. Since each plant contains many thousands of genes, and the breeder is seeking to combine a range of traits in one plant, such as high yield, quality and resistance to disease, developing a successful variety is an extremely lengthy process–up to 12 years in the case of cereals, even longer for potatoes. Plant breeding has been compared to playing a fruit machine–not with three reels, but several hundred. The skill of the plant breeder lies in improving his chances of hitting the jackpot by combining all the desired characteristics in the same variety.

PLANT GENETICS

Knowledge of population genetics, quantitative genetics, probability theory and statistics is indispensable for understanding equilibria and shifts with regard to the genotypic composition of a population, its mean value and its variation. The subject of population genetics is the study of equilibria and shifts of allele and genotype frequencies in populations.

These equilibria and shifts are determined by five forces:

- Mode of reproduction of the considered crop

The mode of reproduction is of utmost importance with regard to the breeding of any particular crop and the maintenance of already available varieties. This applies both to the natural mode of reproduction of the crop and to enforced modes of reproduction, like those applied when producing a hybrid variety. In plant breeding theory, crops are therefore classified into the following categories: cross-fertilizing crops, self-fertilizing crops, crops with both cross- and self-fertilization and asexually reproducing crops. It is explained that even within a specific population, traits may differ with regard to their mode of reproduction.

- Selection
- Mutation
- Immigration of plants or pollen, *i.e.* immigration of alleles
- Random variation of allele frequencies

A population is a group of (potentially) interbreeding plants occurring in a certain area, or a group of plants originating from one or more common ancestors. The former situation refers to cross-fertilizing crops (in which case the term Mendelian population is sometimes used), while the latter group concerns, in particular, self-fertilizing crops. In the absence of immigration the population is said to be a closed population.

Examples of closed populations are:

- A group of plants belonging to a cross-fertilizing crop, grown in an isolated field, *e.g.* maize or rye (both pollinated by wind), or turnips or Brussels sprouts (both pollinated by insects)
- A collection of lines of a self-fertilizing crop, which have a common origin, *e.g.* a single-cross, a three-way cross, a backcross

The subject of quantitative genetics concerns the study of the effects of alleles and genotypes and of their interaction with environmental conditions. Population genetics is usually concerned with the probability distribution of genotypes within a population (genotypic composition), while quantitative genetics considers phenotypic values (and statistical parameters dealing with them, especially mean and variance) for the trait under investigation.

In fact population genetics and quantitative genetics are applications of probability theory in genetics. An important subject is, consequently, the derivation of probability distributions of genotypes and the derivation of expected genotypic values and of variances of genotypic values. Generally, statistical analyses comprise estimation of parameters and hypothesis testing. In quantitative genetics statistics is applied in a number of ways. It begins when considering the experimental design to be used for comparing entries in the breeding programme.

Considered across the entries constituting a population (plants, clones, lines, families) the expression of an observed trait is a random variable. If the expression is represented by a numerical value the variable is generally termed phenotypic value, represented by the symbol p.

Two genetic causes for variation in the expression of a trait are distinguished. Variation controlled by so-called major genes, *i.e.* alleles that exert a readily traceable effect on the expression of the trait, is called qualitative variation. Variation controlled by so-called polygenes, *i.e.* alleles whose individual effects on a trait are small in comparison with the total variation, is called quantitative variation. In Note: it is elaborated that this classification does not perfectly coincide with the distinction between qualitative traits and quantitative traits.

The former paragraph suggests that the term *gene* and *allele* are synonyms. According to Rieger, Michaelis and Green a gene is a continuous region of DNA, corresponding to one (or more) transcription units and consisting of a particular sequence of nucleotides. Alternative forms of a particular gene are referred to as alleles. In this respect the two terms 'gene' and 'allele' are sometimes interchanged. Thus the term 'gene frequency' is often used instead of the term 'allele frequency'.

The term locus refers to the site, alongside a chromosome, of the gene/ allele. Since the term 'gene' is often used as a synonym of the term 'locus', we have tried to avoid confusion by preferential use of the terms 'locus' and 'allele' (as a synonym of the word gene) where possible. In the case of qualitative variation, the phenotypic value p of an entry (plant, line, family) belonging to a genetically heterogeneous population is a discrete random variable. The phenotype is then exclusively (or to a largely traceable degree) a function f of the genotype, which is also a random variable G.

Thus,

$$p = f(G)$$

It is often desired to deduce the genotype from the phenotype. This is possible with greater or lesser correctness, depending for example on the degree of dominance and sometimes also on the effect of the growing conditions on the phenotype. A knowledge of population genetics suffices for an insight into the dynamics of the genotypic composition of a population with regard to a trait with qualitative variation: application of quantitative genetics is then superfluous.

Note: All traits can show both qualitative and quantitative variation. Culm length in cereals, for instance, is controlled by dwarfing genes with major effects, as well as by polygenes. The commonly used distinction between qualitative traits and quantitative traits is thus, strictly speaking, incorrect. When exclusively considering qualitative variation, *e.g.* with regard to the traits in pea (*Pisum sativum*) studied by Mendel, this book describes the involved trait as a trait showing qualitative variation.

On the other hand, with regard to traits where quantitative variation dominates – and which are consequently mainly discussed in terms of this variation – one should realise that they can also show qualitative variation.

In this sense the following economically important traits are often considered to be 'quantitative characters':

- Biomass
- Yield with regard to a desired plant product
- Content of a desired chemical compound (oil, starch, sugar, protein, lysine) or an undesired compound
- Resistance, including components of partial resistance, against biotic or abiotic stress factors
- Plant height

In the case of quantitative variation p results from the interaction of a complex genotype, *i.e.* several to many loci are involved, and the specific growing conditions are important. In this book, by complex genotype we mean the sum of the genetic constitutions of all loci affecting the expression of the considered trait. These loci may comprise loci with minor genes (or polygenes), as well as loci with major genes, as well as loci with both. With regard to a trait showing quantitative variation, it is impossible to classify individual plants, belonging to a genetically heterogeneous population, according to their genotypes. This is due to the number of loci involved and the complicating effect on p of (some) variation in the quality of the growing conditions. It is, thus, impossible to determine the number of plants representing a specified complex genotype. (With regard to the expression of qualitative variation this may be possible!). Knowledge of both population genetics and quantitative genetics is therefore required for an insight into the inheritance of a trait with quantitative variation.

The phenotypic value for a quantitative trait is a continuous random variable and so one may write:

$$p = f(G,e)$$

Thus the phenotypic value is a function f of both the complex genotype (represented by G) and the quality of the growing conditions (say environment, represented by e). Even in the case of a genetically homogeneous group of plants (a clone, a pure line, a single-cross hybrid) p is a continuous random variable.

The genotype is a constant and one should then write:

$$p = f(G,e)$$

Regularly in this book, simplifying assumptions will be made when developing quantitative genetic theory.

Especially the following assumptions will often be made:

- Absence of linkage of the loci controlling the studied trait(s)
- Absence of epistatic effects of the loci involved in complex genotypes.

These assumptions will now be considered.

Absence of Linkage

The assumption of absence of linkage for the loci controlling the trait of interest, *i.e.* the assumption of independent segregation, may be questionable in specific cases, but as a generalisation it can be justified by the following reasoning. Suppose that each of the n chromosomes in the genome contains M loci affecting the considered trait. This implies presence of n groups of,

$$\binom{M}{2}-$$

pairs of loci consisting of loci which are more strongly or more weakly linked. The proportion of pairs consisting of linked loci among all pairs of loci amounts then to,

$$\frac{n\binom{M}{2}}{\binom{nM}{2}}=\frac{n.M!}{2!(M-2)!}\times\frac{2!(nM-2)!}{(nM)!}=\frac{M-1}{nM-1}=\frac{1-\dfrac{1}{M}}{n-\dfrac{1}{M}}$$

For $M = 1$ this proportion is 0; for $M = 2$ it amounts to 0.077 for rye (*Secale cereale*, with $n = 7$) and to 0.024 for wheat (*Triticum aestivum*, with $n = 21$); for $M = 3$ it amounts to 0.100 for rye and to 0.032 for wheat. For $M \to \infty$ the proportion is 1 n; *i.e.* 0.142 for rye and 0.048 for wheat.

One may suppose that loci located on the same chromosome, but on different sides of the centromere, behave as unlinked loci. If each of the n chromosomes contains,

$$m\left(=\frac{1}{2}M\right)$$

relevant loci on each of the two arms then there are $2n$ groups of,

$$\binom{m}{2}-$$

pairs consisting of linked loci. Thus considered, the proportion of pairs consisting of linked loci amounts to,

$$\frac{2n\binom{m}{2}}{\binom{2nm}{2}}=\frac{2n.m!}{2!(m-2)!}\times\frac{2!(2nm-2)!}{(2nm)!}=\frac{1-\dfrac{1}{m}}{2n-\dfrac{1}{m}}$$

For $m = 1$ this proportion is 0; for $m = 2$ it amounts to 0.037 for rye and to 0.012 for wheat; for $m = 3$ it amounts to 0.049 for rye and to 0.016 for wheat. For $m \to \infty$ the proportion is,

$$\frac{1}{2n}$$.*e.* 0.071 for rye and 0.024 for wheat.

For the case of an even distribution across all chromosomes of the polygenic loci affecting the considered trait it is concluded that the proportion of pairs of linked loci tends to be low. (In an autotetraploid crop the chromosome number amounts to $2n = 4x$. The reader might like to consider what this implies for the above expressions.)

GENETIC ENGINEERING

Genetically modified plants are created by the process of genetic engineering, which allows scientists to move genetic material between organisms with the aim of changing their characteristics. All organisms are composed of cells that contain the DNA molecule. Molecules of DNA form units of genetic information, known as genes. Each organism has a genetic blueprint made up of DNA that determines the regulatory functions of its cells and thus the characteristics that make it unique. Prior to genetic engineering, the exchange of DNA material was possible only between individual organisms of the same species. With the advent of genetic engineering in 1972, scientists have been able to identify specific genes associated with desirable traits in one organism and transfer those genes across species boundaries into another organism. A gene from bacteria, virus, or animal may be transferred into plants to produce genetically modified plants having changed characteristics. Thus, this method allows mixing of the genetic material among species that cannot otherwise breed naturally. The success of a genetically improved plant depends on the ability to grow single modified cells into whole plants. Some plants like potato and tomato grow easily from single cell or plant tissue. Others such as corn, soy bean, and wheat are more difficult to grow.

After years of research, plant specialists have been able to apply their knowledge of genetics to improve various crops such as corn, potato, and cotton. They have to be careful to ensure that the basic characteristics of these new plants are the same as the traditional ones, except for the addition of the improved traits. The world of biotechnology has always moved fast, and now it is moving even faster. More traits are emerging; more land than ever before is being planted with genetically modified varieties of an ever-expanding number of crops. Research efforts are being made to genetically modify most plants with a high economic value such as cereals, fruits, vegetables, and floriculture and horticulture species.

BIOLOGICAL CELL

It only takes one biological cell to create an organism. In fact, there are countless species of single celled organisms, and indeed multi-cellular organisms like ourselves. A single cell is able to keep itself functional by owning a series of '*miniature machines*' known as *organelles*. The following list looks at some of these organelles and other characteristics typical of a fully functioning cell.

- *Mitochondrion:* An important cell organelle involved in respiration
- *Cytoplasm:* A fluid surrounding the contents of a cell and forms a vacuole
- *Golgi Apparatus:* The processing area for the creation of a glycoprotein
- *Endoplasmic Reticulum:* An important organelle heavily involved in protein synthesis.
- *Vesicles: Packages* of substances that are to be used in the cell or secreted by it.
- *Nucleus*: The "brain" of a cell containing genetic information that determines every natural process within an organism.
- *Cell Membrane:* Also known as a plasma membrane, this outer layer of a cell assists in the movement of molecules in and out the cell plays both a structural and protective role
- *Lysosomes:* Membranous sacs that contain digestive enzymes

CELL WALL

A structure that characteristically is found in plants and prokaryotes and not animals that plays a structural and protective role.

CELL SPECIALISATION

Cells can become specialised to perform a particular function within an organism, usually as part of a larger tissue consisting of many of the same cells working in tandem:

- *Nerve cells* to operate as part of the nervous system to send messages back and forth via the brain at the centre of the nerve system.
- Skin cells for waterproof protection and protection against pathogens in the open air environment.
- *Xylem* tubes to transport water around plants and to provide structural support for the plant as a whole.

Cells combine their efforts in these tissue types to perform a common cause. The task of the specialised cell will determine in what way it is going to be specialised, because different cells are suited to different purposes, as illustrated in the above list:

- Muscle cells are long and smooth in structure and their elastic nature allows these cells to perform flexible movements, just as they do in our own body's.
- Some *white blood cells* contain powerful digestive enzymes to eliminate pathogens by breaking them down to the molecular level.
- Cells at the back of the eye are sensitive to light stimuli, and thus can *interpret* differences in light intensity which can in turn be interpreted by our nervous system and brain.

Many of these cells contain organelles, though after some cells are specialised, they do not possess particular characteristics as they do not require them to be there. *i.e.* efficiency is the key, no resources are wasted and the resources available are put to their idyllic optimum.

THE CELL MEMBRANE

The *cell membrane,* otherwise known as the plasma membrane is a semi-permeable structure consisting mainly of *phospholipid* (fat) molecules and proteins. They are structured in a *fluid mosaic model,* where a double layer of phospholipid molecules provide a barrier accompanied by proteins. It is present round the circumference of a cell to acts as a barrier, keeping foreign entities out the cell and its contents (like cytoplasm) firmly inside the cell. The plasma membrane allows only selected materials to pass in and out of a cell, and is thus known as a selectively permeable membrane.

CELL TRANSPORT

There are three methods in which ions are transported through the cell membrane into the cell,

- *Active Transport*: Active transport is the transport of molecules with the active assistance of a carrier that can transport the material against a natural *concentration gradient.*
- *Passive Transport (Diffusion):* The movement of molecules from areas of high concentration (*i.e.* outside a cell) to areas of low concentration (*i.e.* within a cell) via a carrier. This process does not require energy.
- *Simple Diffusion:* The movement of molecules from areas of high concentration to areas of low concentration in a free state. *Osmosis* of water involves this type of diffusion through a selectively permeable membrane (*i.e.* plasma membrane)

THE BREAKDOWN OF MATERIALS IN A CELL

In cells, sometimes it is required to breakdown more complex molecules into more simple molecules, which can then be 're-built' into what is needed by the body with these new raw materials. '*Pinocytosis*' where to contents of a structure (such as bacteria) are *drank,* essentially by breaking down molecules into a drinkable form. '*Phagocytosis*' where contents are 'eaten'.

ABSORPTION AND SECRETION

Absorption is the uptake of materials from a cells' external environment. Secretion is the ejection of material.

BIOLOGICAL ENERGY-ADP AND ATP

ATP stands for Adenosine Tri-Phosphate, and is the energy used by an organism in its daily operations. It consists of an *adenosine* molecule and three inorganic *phosphates*. After a simple reaction breaking down ATP to *ADP*, the energy released from the breaking of a molecular bond is the energy we use to keep ourselves alive.

ATP TO ADP-ENERGY RELEASE

This is done by a simple process, in which one of the phosphate molecules is broken off, therefore reducing the ATP from 3 phosphates to 2, forming ADP (Adenosine Diphosphate after removing one of the phosphates {Pi}). This is commonly wrote as ADP + Pi. When the bond connecting the phosphate is broken, *energy* is released. While ATP is constantly being used up by the body in its biological processes, the energy supply can be bolstered by new sources of glucose being made available via eating food which is then broken down by the digestive system to smaller particles that can be utilised by the body. On top of this, ADP is built back up into ATP so that it can be used again in its more energetic state. Although this conversion requires energy, the process produces a net gain in energy, meaning that more energy is available by re-using ADP+Pi back into ATP.

GLUCOSE AND ATP

Many ATP are needed every second by a cell, so ATP is created inside them due to the demand, and the fact that organisms like ourselves are made up of millions of cells.

Glucose, a sugar that is delivered via the bloodstream, is the product of the food you eat, and this is the molecule that is used to create ATP. Sweet foods provide a rich source of readily available glucose while other foods provide the materials needed to create glucose. This glucose is broken down in a series of *enzyme* controlled steps that allow the release of energy to be used by the organism. This process is called respiration.

RESPIRATION AND THE CREATION OF ATP

ATP is created via respiration in both animals and plants. The difference with plants is the fact they attain their food from elsewhere. In essence, materials are harnessed to create ATP for biological processes. The energy can be created via cell respiration.

The process of respiration occurs in 3 steps (when oxygen is present:

- Glycolysis
- The Kreb's Cycle
- The Cytochrome System

CELL RESPIRATION

As mentioned in the previous page on ATP, the process of *respiration* is split into 3 distinct areas that occur at different parts of the cell. Respiration involves the *oxidation* of foodstuff (*i.e.* glucose) in order to create ATP.

Respiration can occur with or without oxygen, *aerobic* and *anaerobic* respiration respectively.

GLYCOLYSIS

Glycolysis occurs in the *cytoplasm* of a cell where a 6 carbon glucose molecule (the broken down food that you ate earlier) is broken down by enzymes into a 3 carbon *pyruvic acid*. The execution of this process requires 2 ATP, and produces a net gain of 2 ATP. The enzymes involved remove hydrogen from the glucose (oxidation) where they take these hydrogen atoms to the cytochrome system, explained soon.

In anaerobic respiration, this is where the process ends, glucose is split into 2 molecules of pyruvic acid. When oxygen is present, pyruvic is broken down into other carbon compounds in the Kreb's Cycle. When it is not present, the pyruvic acid is broken down into lactic acid (or carbon dioxide and ethanol).

THE KREB'S CYCLE

When oxygen is present, respiration can harness more ATP from a single unit of glucose. The *pyruvic acid* from the *glycolysis* stage diffuses into a cell organelle called a mitochondrion (pl. *mitochondria*). These mitochondria are sausage shaped structures that host a large surface area for the respiration to occur on. The pyruvic acid is then subject to more enzymes which break it down into a 2 carbon compound, as seen below.

The diagram illustrates the Kreb's cycle, consisting of three main actions:

- The carbon element is in an infinite cycle where the 2 carbon compound derived from pyruvic acid binds with the 4 carbon compound that is always present in the cycle.
- CO_2 is released, where the oxygen that is present in *aerobic respiration* combines with carbon from the carbon compounds which is released as CO_2. Hence the need for animals to breath out and expel this CO_2.
- Enzymes oxidize the carbon compounds and transport the hydrogen atoms to the cytochrome system.

THE CYTOCHROME SYSTEM

The cytochrome system, also known as the hydrogen carrier system (or the *electron transport system*) are where the reduced hydrogen carriers transport hydrogen atoms from the glycolysis and Kreb's cycle stages. The cytochrome system is found in the many *cristae* of mitochondria, which are tiny stalked particles found on its outer layer.

The system contains many 'hydrogen acceptors' which hydrogen can be added to. By following the path of a hydrogen atom, we can see how the cytochrome system works:

- Some *coenzymes* from earlier stages are transferred to the next coenzymes.
- B is then oxidised, therefore the coenzyme releases the hydrogen and energy is made available.
- The released hydrogen atom binds with 2 oxygen atoms (oxygen is available in aerobic respiration) which produces water, a by-product of respiration.

The diagram illustrates this flow of hydrogen within the cytochrome system and how energy is made available by the flow of these atoms. The green circles illustrate where energy is made available via oxidation. Overall their is a gain of 38 ATP from one molecule of glucose in aerobic respiration. The food that we eat provides glucose required in respiration. In plants, energy is also acquired via respiration, but the mechanism of delivering glucose to the respiration process is a little different.

GENETIC ENGINEERING TECHNOLOGY

Genetic manipulation for biotechnological plant improvement relies on the ability to modify genes as DNA sequences. This requires utilization of standard, well-documented techniques. Enzymes are the basic tool kit of the genetic engineer, and can be used for basic functions of manipulating nucleic acids in such a way as to copy (DNA polymerase), transcribe (RNA polymerase), cut at specific sites (restriction endonucleases), cut and degrade in a nonspecific nature (nucleases) and join pieces together (DNA ligase).

DNA libraries are another important resource for genetic manipulation as they represent a means of storing and screening sequences from an entire organism. The two basic types of library are the cDNA library and the genomic library. A cDNA (complementary DNA) library is based on expressed sequences. Such a library is made by extracting mRNA from a particular plant tissue of interest, then making first single-stranded DNA (via reverse transcriptase) then double-stranded DNA (via DNA polymerase). A cDNA library will then represent only those genes which are expressed in the tissue of interest (or in response to a stress of some sort), and will be devoid of introns and regulatory sequences.

A genomic library represents the entire genome (often including the organellar genomes), and will include all sequences whether they are expressed in the donor tissue or not, and even sequences which are not expressed at all. Genomic libraries will therefore include highly repeated sequences such as rRNA genes and nonexpressed simple sequence repeats (SSRs), as well as genes and their regulatory sequences and introns.

Genomic libraries may be cloned and stored in lambda bacteriophage, cosmids or larger fragments which will often include multiple genes in bacterial artificial chromosomes (BACs) and yeast artificial chromosomes

(YACs). BAC library inserts may average 150 kb, whereas YAC inserts are larger on average, at up to 1 Mb of DNA.

Overall, the ability to genetically engineer a plant relies on:

- A genetic transformation system for transferring DNA into plant cells.
- A plant regeneration system to produce plants from the transgenic cells.
- A gene construct which can modify the plant phenotype in a desirable and controlled manner.

GENETIC TRANSFORMATION TECHNIQUES

A number of genetic transformation techniques have been developed since the first report of a transgenic plant. With few exceptions, these methods rely on the ability to transfer DNA to a single cell or small group of cells, such as nondifferentiated leaf tissue or meristems. Then some form of tissue culture is required to regenerate whole, fertile transgenic plants for the process to be completed. Therefore, the first requirement for plant transformation is a reliable plant regeneration system using explants/tissues which are amenable to the DNA delivery system. This requirement was, more than any other factor, the greatest impediment to genetic engineering of members of the cereal and grain legume families – the two most important crop groups. Regeneration in most of these species is difficult, particularly for species like soybean and chickpea.

At best, regeneration is extremely genotype-dependent for species such as rice, where the *japonica* rices are easier to manipulate *in vitro* than the *indica* rices, which are more important on a worldwide basis. However, international effort in most of the important cereal species has largely overcome the problem, although there are still genotypes of a number of species which remain recalcitrant to regeneration. The cereals and pulses (grain legumes) proved somewhat recalcitrant to genetic transformation until the late 1980s, when rice became the first cereal to become reliably transformed at high frequency, with soybean first successfully transformed via *Agrobacterium* in 1988, although repeatable transformation was only achieved later using a microprojectile approach. There are three basic transformation systems being routinely used in most genetic engineering laboratories internationally.

These are:

- *Agrobacterium*-mediated transformation.
- DNA uptake via protoplasts.
- Microprojectile bombardment.

Agrobacterium-mediated Transformation

Agrobacterium tumefaciens, the causal agent of crown gall disease, transfers a portion of its own DNA (T-DNA) to the plant, where it is stably incorporated into the plant genome. The T-DNA is delimited by specific 25 bp repeat sequences, known as the border sequences. None of the T-DNA is actually required for transfer, and whatever DNA is bound by the borders will be transferred, hence the pathogenic DNA can be effectively replaced with genes of interest, allowing delivery of these desirable genes to cells from which whole plants can be regenerated.

The ability of *Agrobacterium* to transfer genes to plants is known as virulence, and is largely controlled by the *vir* genes on the Ti plasmid, and to a lesser extent, the chromosomal virulence (*chv*) genes. The *vir* genes are induced by the plant's wound response, in particular the production of phenolic compounds such as acetosyringone.

The Vir proteins then facilitate the identification, copying, transfer and targeting to the host nucleus, and incorporation on a plant chromosome of the T-DNA. Generally the simplest manner by which to replace the T-DNA is to remove it, leaving a 'disarmed' Ti plasmid with the *vir* genes, then introducing a binary plasmid which contains the gene(s) of interest within the border sequences. Hence, with the induction of virulence, the Vir proteins will effect the transfer of the engineered T-DNA.

Most dicotyledonous species are hosts to *Agrobacterium* and as such can be genetically engineered in this manner. *Agrobacterium*-mediated transformation remains the method of choice for most species of the *Solanaceae* and *Brassicae,* and other top 20 most important plants such as grape, orange, cotton, sugarbeet and apple. While there are a number of reports of successful transformation of important grain legumes including soybean, chickpea and *Phaseolus* bean, there is generally great genotypedependence of such methods. It has been shown repeatedly that plant genotype 3 *Agrobacterium* strain interactions can vary from complete absence of transformation to 100% transformation within many species, including soybean, pea and sugarbeet.

Most important monocotyledonous species were once regarded as being outside the host range of *Agrobacterium.* In the late 1980s it became evident that some monocot species, including asparagus, yam and onion, were transformable using *Agrobacterium.*

Early reports of *Agrobacterium*-mediated transformation of rice, maize and sorghum remained largely unrepeatable in these and other labs, until the first report of a high-frequency, repeatable, and painstakingly substantiated system was published for three *japonica* rice cultivars Chan *et al*., (1993) first reported stable transformation of rice using *A. tumefaciens*.

Progeny analysis and Southern hybridization demonstrated that it was possible to generate transgenic rice via *Agrobacterium,* albeit at low frequency. Hiei *et al.* (1994) improved the efficiency greatly by co-cultivating callus originating from scutella of immature embryos with a super binary vector then selecting on hygromycin, and produced over 400 primary transgenics of three cultivars.

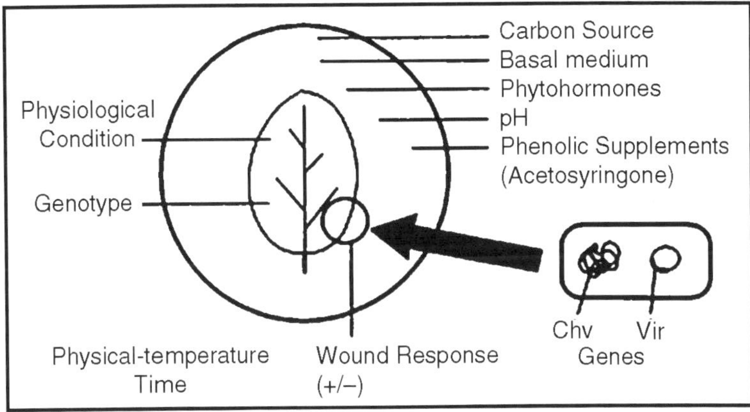

Fig. A Summary of the Plant,Co-cultivation Medium,Bacterial and Physical Factors which Affect the Efficiency of Virulence Induction and Arobacterium-mediated Gene Transfer.

Detailed analyses of 29 plants and their progeny revealed that stable transformation was possible, and in most cases marker genes segregated in the expected Mendelian fashion.

In addition, these authors sequenced T-DNA/plant DNA junctions and found that incorporation occurred at similar sites to those found in transgenic tobacco plants. The 'super binary' plasmid contained *virB* and *virG* genes from the hypervirulent pTiBo542 plasmid, and, although it enhanced transformation frequencies, was not an absolute requirement for transformation. Basic requirements for *Agrobacterium*-mediated transformation are host recognition, virulence gene induction, the ability to select and proliferate transgenic cells, and the ability to regenerate whole, fertile plants from these cells. Host recognition and virulence gene induction require the correct carbon source and pH, and the presence of virulence-inducing molecules. Many wounded plant tissues produce signal molecules such as acetosyringone, which set in train the events of T-DNA transfer and integration into the host nuclear genome.

Protoplasts

Direct gene transfer has been the method of choice for monocotyledons, predominantly because most of these species have proved recalcitrant to *Agrobacterium*-mediated transformation. Protoplasts are cells with the cell wall enzymatically removed. In the absence of a cell wall, macromolecules such as proteins and nucleic acids may be taken up by such cells. The efficiency of uptake of DNA is greatly enhanced by the application of physiological shock, such as an ionic, osmotic or electrical current, which causes reversible permeabilization of the cell membrane. DNA has been shown to transfer to cells of most major cereals using the electric shock, or electroporation, methodology. If the DNA enters the nucleus and is stably incorporated into the plant genome, the cell is transgenic. This has been achieved with all major cereals (e.g. sorghum). One of the advantages of this methodology is that a large number of theoretically competent cells can be exposed to DNA at one time.

Regeneration of whole, fertile plants from cereal protoplasts has remained the greatest hurdle to electroporation-mediated transformation. Even when protoplasts are regenerable and DNA transfer is effected, it is difficult to combine the two. The physiological shock necessary for DNA uptake may often be enough to prevent subsequent cell wall synthesis or substantially reduce regenerability. As protoplast cultures usually require greater time *in vitro*, there may also be an increased frequency of somaclonal variation. Success has been achieved with regeneration of rice, wheat and maize but protoplasts remain a widely used method for rice only, with microprojectile-mediated transformation overtaking it in efficiency.

Microprojectile Bombardment

The potential for genetic transformation with microprojectiles was first reported by Klein *et al.* (1987), who achieved strong transient expression of a reporter gene in intact onion tissues. The first report of fertile transgenic plants and their progeny described stably transformed tobacco plants with Mendelian inheritance among selfed progenies. However, much of the focus with microprojectile bombardment has concentrated on cereals, due to the apparent inability of *Agrobacterium* to transfer genes to most cereal species, and the recalcitrance and genotype-specificity of cereals (with the exception of rice) to regeneration of fertile plants from transformed protoplasts.

The microprojectile bombardment, or 'Biolistic' (commercial description used by Bio Rad Labs), approach is a simple concept, based on the introduction of DNA-coated small inert projectiles into plant cells, not unlike the action of a microscale shotgun. The system is a modification of one used by virologists, whereby high-velocity microprojectiles are used to facilitate viral infection. To adapt this for transformation, the major alterations required were to reduce microprojectile size to reduce the extent of cell damage, to be able to perform the procedure under sterile conditions to exclude contamination by bacteria or fungi, and most importantly, to ensure that the target tissue be regenerable to produce whole, fertile plants.

Here lies one of the greatest advantages of the microprojectile approach over protoplast and *Agrobacterium*-mediated techniques. Effectively, any target tissue may be used for bombardment, so the most regenerable tissues can be used. In addition, there does not appear to be any genotype specificity in the ability to deliver DNA, which is demonstrably the case with *Agrobacterium*mediated transformation (e.g. Owens and Smigocki, 1988). The genotype dependence of plant regeneration remains the greatest hurdle in most species, although this is by no means the barrier that protoplast regeneration presents.

The microprojectiles, which are commonly an inert metal such as tungsten or gold, range in diameter from 1 to 4 mm; a velocity of approximately 250 m s-1 is required to penetrate cell walls and membranes. For successful stable transformation, the DNA must be delivered intact to the nucleus where the DNA will then be incorporated onto a chromosome. This requires that the velocity of microprojectiles be sufficient to penetrate cells, but not cause excessive damage.

For most of the cereals, the target tissue of choice is the immature zygotic embryo, particularly the scutellum, or embryogenic callus cultures, usually derived from zygotic embryos. Other targets such as embryogenic suspension cultures have also been used. As can be seen, the most important cereal species have been recently transformed using this approach. In addition, transgenic oats and rye have been produced with microprojectiles.

Early microprojectile apparatus was based on a gunpowder charge for particle acceleration. Improvements in safety and reliability were made by using electrical discharge or helium gas to provide acceleration, and most laboratories currently use helium-based guns, such as the commercially available Biolistic gun.

Tissue sterility is maintained by performing all operations in a laminar flow cabinet with bombardment actually taking place in a sealed chamber under a slight vacuum. One of the drawbacks of the microprojectile approach is the cost of the Biolistic gun, hence other laboratories have developed simpler apparatus, such as the particle inflow gun (PIG), which relies on a solenoid to regulate helium gas inflow. The great advantage of the PIG is that it can be built for around US$1000. There are potential questions about the use of such apparatus and whether there is any contravention of international patents.

Alternative Transformation Methods

While the great majority of transgenic plants worldwide are being produced via *Agrobacterium*, microprojectile bombardment and, to a lesser extent, protoplasts, a number of alternative transformation methods have been trialled. Although not intended to be an inclusive list, these include:

- Virus-mediated transformation.
- Pollen pathway or intact inflorescence methods.
- Microinjection.
- Silicon carbide fibres.
- Electroporation of intact tissues.
- Electrophoresis of intact tissues.

While it is not possible to discuss the success, advantages and disadvantages of these methods, the subject has been recently reviewed by Songstad *et al.* (1995).

The pollen pathway has long been recognized as a potentially useful means of introducing genes to the germline, with its greatest advantage being the avoidance of tissue culture and the need to regenerate undifferentiated tissues such as callus or protoplasts.

The absence of any time in plant tissue culture should also totally avoid the problem of somaclonal variation. There have been many difficulties with this approach, however, and there are no substantiated instances of genetic transformation via the pollen pathway. More promising methods are the use of silicon carbide fibres and electroporation/ electrophoresis of intact tissues. Stable tissue transformation of tobacco and corn (maize) suspension cultures has been achieved following agitation with silicon carbide fibres. Stable cereal transformants have been produced by electroporating germinating rice seeds and immature embryos of corn, and by electrophoresis of germinating barley seeds. However, until some of these alternative systems can be made more reliable and efficient, it would appear that the two most widely used transformation systems will remain *Agrobacterium* and, in the case of cereals, microprojectile bombardment.

HOW TO FACE THE NEW CHALLENGE OF GENE TECHNOLOGY

With few exceptions classical bioprospecting for drugs has not proved economically beneficial for developing nations, but nor has it directly damaged these economies. In contrast, bioprospecting for genes may soon pose a real threat to the economic survival of these biodiversity-rich countries. The farmers of the North are currently suffering under low prices and overproduction of certain traditional crops and livestock. As a result they are looking for new markets and products.

Modern biotechnology promises to aid Northern farmers in this endeavour, eliminating or displacing traditional export commodities from developing countries and transferring production or substitutes from the farm fields of the South to those of the North. Quite possibly the transfer may even skip the Northern farms and jump straight into bioreactors. In Africa alone, US$ 10 billion in exports are vulnerable to industry-induced changes in raw material prices and requirements. Most developments in plant biotechnology have been achieved with crops cultivated mainly in industrialized countries. This denies the farmers of developing countries the chance to benefit from the new agricultural opportunities of gene technology.

As mentioned above, Calgene's high-laurate canola oil may displace coconut and palm kernel oil, posing a threat to the economic survival of millions of farm families in the South. In the Golfito region of Costa Rica the government has started a programme to grow oil palms on banana fields deserted by the US fruit multinationals. All these efforts, which are partially financed with Northern developmental aid, may vanish into thin air as a result of Calgene's new rape seed.

Thaumatin, a sweet-tasting basic protein from the tropical plant *Thaumatococcus,* has been traditionally used in West Africa as a sweetener. With the collection of *Thaumatococcus* fruits for the British food industry, the population in this region earned a large part of its income.

However, the thaumatin gene has now been cloned and the sweetening protein can be produced by large-scale fermentation of brewer's yeast at low cost. The same applies to natural compounds like indigo, which can now be produced by the fermentation of *Escherichia coli* engineered with genes from a toluol-degrading subspecies of the soil bacterium *Pseudomonas putida.*

Products like vanilla, pyrethrum and rubber may follow this same path. Along the same lines, the extension of patent laws to the developing nations through the GATT could mean that the biotech industry obtains a monopoly on genetically engineered livestock and crops, which farmers in developing countries must cultivate under constraint in order to remain competitive. For example, US-Patent No. 5 159 135 of Agracetus covers all genetically engineered cotton. This patent is a warning of potential future problems, and has already caused an outcry in developing countries like India. If the biotech industry continues developing without adequate controls, the consumers and farmers of developing countries may even end up paying royalties on biotech products that were originally developed from their very own resources and knowledge.

As a step in the right direction, the US Patent and Trademark Office reversed its decision to grant the Agracetus patent at the end of 1994, but primarily as a result of pressure from the US biotech industry which argued that patents like this will inhibit research and development. Such issues bring important questions to light: How can developing nations be motivated to conserve their biodiversity under these threatening circumstances?

Are the regulations and tools of modern bioprospecting, as described above, sufficient to face this challenge? The Biodiversity Convention attempts to address this new threat by requiring that access to biodiversity's genetic potential be combined with the biotechnology transfer to the South in order for those countries to develop their own methods of sustainable biodiversity utilization. Nevertheless, the Convention suffers from three major drawbacks.

- The treaty was not ratified by the USA, the leading country in biotechnological research, development and application.
- It specifically excludes (under US pressure) *ex situ* genebank material collected before the enactment of the treaty. As a result, huge stocks of germplasm collected by the North, mostly in tropical and subtropical countries, are not restricted by the Convention. The recent transfer by the Consultative Group on International Agricultural Research of its 12 genebanks to the auspices of the UN must be the first step in keeping the access of developing countries open to their own resources.

- It still remains nearly impossible to control the illegal transfer of genetic material into the North. Genes can be cloned from minute amounts of DNA or RNA and isolated from biological material that easily fits into an airmail envelope. Genes do not have tags designating their country of origin, and once they are cloned, they are no longer controlled by their source country. This is quite different from the isolation of natural compounds from plants, where larger amounts of plant material must be collected and, for the process of isolation and structure elucidation, must be re-collected. In this last case, controlling the flow of biological material is possible simply because industry will eventually require sample resupply at a given point, and for this the industry needs reliable partners in countries of origin.

These issues must be approached in an active and more aggressive manner than traditionally used. During collaborations with traditional pharmaceutical and biotech companies like Merck & Co., Bristol-Myers Squibb and the British Technology Group, INBio used such an approach, proving to industry that fair partnerships are mandatory and conducive to success. More importantly, INBio demonstrated that reliable applied research is possible in a developing country, and that technology transfer to acquire necessary know-how and equipment works to the advantage of the industrial partner as well. The same applies to the biotech industry.

In this case, the transfer of gene technology is not critical, in contrast to natural compound chemistry, because it does not require million-dollar investments in laboratory infrastructure. Rather, the industry relies on the know-how already existing in many source countries. Gene technology also represents a promising development opportunity for countries that do not have large research budgets at their disposal. Moreover, gene technology is a very straightforward approach, relying on natural history observations (e.g. whether certain plants show natural resistance to pathogens), and not involving the automated random screening of thousands of samples. Biodiversity inventories, which are already in place in countries like Costa Rica, are a reasonable and advantageous prerequisite for successful 'gene prospecting'.

Costa Rica's aggressive strategy to foster collaborations with the international industry and academic institutions in drug research, gene technology and agriculture actively seeks to develop and patent natural compounds, proteins and genes in Costa Rica based on a foundation of national research. Collaborations of this nature will help launch Costa Rica onto a scientific and technological plane that offers services and goods that are both competitive and compatible with those of industrial nations. Simultaneously, INBio works within this strategy to increase knowledge about Costa Rican biodiversity in general, access revenues for further conservation efforts, and to assign biodiversity a higher value than it has had in the past. There is little doubt that such activities will encourage society's willingness to preserve biodiversity for future generations, by making it worthwhile for the Costa Rican population to maintain tropical forests and other ecosystems on their own.

GENE MAPPING

An important, though by no means an essential, step in genetic analysis is to produce genetic maps of the marker loci. Such maps are often referred to as 'framework maps' because they provide a framework within which important genes can be located, as well as providing a means of comparing chromosome organization in other closely or distantly related species. There are two stages to mapping. Firstly, to arrange the markers in a linear sequence separated by an appropriate map distance, *i.e.* to construct a linkage map. The second is to relate the linkage maps to particular recognizable chromosomes.

The latter is often the most difficult and generally not essential for breeding or conservation work. We will, therefore, concentrate on the former. Chromosomes contain a single linear molecule of DNA and hence the markers on that chromosome occur at particular positions along that molecule. Typically, each chromosome contains 107 to 108 base pairs (bp), *i.e.* 104 to 105 kilo-base pairs (kbp) of DNA, while a typical structural gene, coding for a polypeptide chain, would be between 1 and 2 kbp long. Assuming only 10% of the genome is coding DNA, then a chromosome probably contains something of the order of 1000 to 10 000 genes. Fortunately, the fact that the chromosome is a linear molecule means that the genes need to be mapped in only one dimension.

Currently, the only useful method of gene mapping, at least as far as breeders and conservationists are concerned, relies on recombination between homologous chromosomes resulting from genetic exchange, which can be seen as chiasmata at diplotene through first metaphase of meiosis. A single chiasma on a particular chromosome results in half the gametes from such a meiosis being recombinant for that chromosome: half the gametes will contain a copy of that chromosome that is part maternal and part paternal in origin, while the remaining gametes will be entirely parental. Very few plant species have a sufficiently low number of clearly recognizable chromosomes that the number of chiasmata on a particular chromosome can be compared in different nuclei.

However, in many species the total number of chiasmata in each diplotene nucleus can be counted, at least in pollen meiosis, and the average number of chiasmata per nucleus calculated. The number and position of chiasmata on a particular pair of homologous chromosomes will vary from nucleus to nucleus; it is as though there are a large number of potential sites of exchange, but that in any given meiosis only a few sites are actually involved. The longer the chromosome the more potential sites there might be. There will invariably be one chiasma and there may be two, three, four or more; a typical set of results is shown in Table A chromosome that has one chiasma on average is said to be 50 centimorgans (cM) long – a map length unit named after the American geneticist Thomas Hunt Morgan.

This relates to the fact mentioned above that one chiasma results in 50% recombinant chromosomes. By extension, a chromosome with an average of 2.5 chiasmata is said to be 125 cM (*i.e.* 2.5 350 cM) long. It appears that, as a rough rule of thumb, each chromosome has on average two chiasmata and so is 100 cM long. It follows, therefore, that providing the haploid chromosome number is known (*n*), the total map length will be approximately 2 3 *n* 3 50 cM. The average number of chiasmata per nucleus has been calculated in many species and this rule generally holds.

Table. Chiasma Fraquency and Mapping.Chiasma Frequency in Chromosomes of Secale Cereale.

Number of chiasmata per chromosome	Frequency	Percentage unrecombined Chromosomes
1	0.267	50.0
2	0.716	25.0
≥3	0.017	12.5
Mean	–	22.0

This rough rule is important because it gives the geneticist an idea of the total genetic map length that should be expected as more genes or markers are mapped. It provides a guide to the extent to which the currently mapped markers cover the full genetic map. Because the map is based in chiasma units, it is not the same as a physical map. The distribution of chiasmata is not uniform over a given chromosome in a species and will vary with the chromosome and the species.

Although a knowledge of the number of chromosomes and, ideally, the mean chiasma frequency per nucleus for the species, indicates the total map length, the actual map has to be constructed from examining the frequencies of progeny in crosses; *i.e.* maps are constructed from information based on the consequences of chiasmata, namely recombination of genetic markers. Recombination of genetic markers is identified by the presence of gametes that contain recombined markers, and such gametes can be recognized from the phenotypes of progeny of a cross.

The frequency of such gametes is an estimate of the recombination frequency between the two markers concerned. The closer two markers are to each other on the chromosome, the less likely a chiasma will occur between them and the less likely they are to recombine. Providing the two markers are sufficiently close that either none or one chiasma occurs between them but never more, then the recombination frequency as a percentage is equal to the map distance in centimorgans as described above. The problem arises when the markers are sufficiently far apart for two or more chiasmata to occur between them in some meioses. It can be shown that not only does a single chiasma between a pair of markers result in 50% recombination between them, but so also do two, three or more chiasmata, on average. It is necessary to say 'on average', because there are a variety of possible consequences with two, three or more chiasmata in any given meiosis but, in practice, when one is looking at the progeny of a cross or self, each progeny will be the result of gametes from different meioses. This fact implies that although map distance increases linearly with the number of chiasmata, the same is not true with the frequency of recombination, which can never be more than 50% for a pair of markers even though they might be 100 cM apart at opposite ends of a chromosome.

These relationships are illustrated in Figure. In order to overcome this problem, mapping functions have been devised to correct for multiple chiasmata. The most common of these are the Haldane (1919) and Kosambi (1943) mapping functions. Haldane's assumes that the probability of none, one, two or more chiasmata in a given interval follows a Poisson distribution, *i.e.* that the chiasmata are independent, random events. Kosambi's method allows for a certain degree of nonindependence in chiasma occurrences. It has long been known by cytogeneticists that chiasma interference occurs over quite large regions of a chromosome.

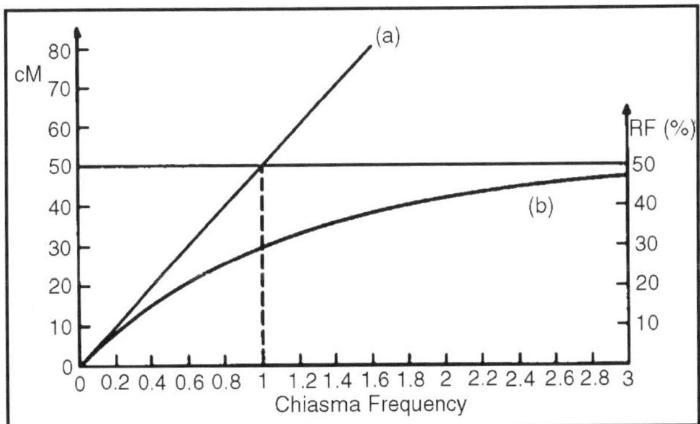

Fig. The Relationship of Chiasma Frequency With (a) Genetic Map Length and(b)Recombination Frequency.

Interference means that if a chiasma occurs at a particular point on the chromosome, the next one will never occur closer than some fixed distance, the interference distance, from it. Beyond this distance, there is a short distance in which the interference disappears and the next chiasma can then occur randomly outside this range. Observational data in various species of plants are confirming this and suggest that the interference distance is generally between 15 and 20 cM, *i.e.* 15 to 20% of a typical chromosome on either side of the chiasma. This has important and useful consequences for genetic and breeding work as will be shown later. Recombination frequencies between pairs of markers can be scored in a wide variety of different crosses. The simplest to use are generations derived from an F1 because only two alleles are segregating, and their distribution in the chromosomes of the parents of the F1 can be determined. The generations that can be used are F2, backcrosses (Bc), recombinant inbred lines (RILs) or doubled haploid (DH) lines, with F2s being the most informative.

General formulae for calculating recombination frequencies are given by Allard. In outbreeding species where F1s are not available, a given individual can be considered as an F1 and its selfed progeny as being an F2, even though the inbred parents do not exist. If it cannot be selfed but controlled crossing to another individual is possible, then again those genes segregating in the cross can be mapped although the situation is more complex. There could be from two to four alleles segregating at each locus and, with respect to any one locus, the cross could represent an F2, Bc1 or Bc2. Moreover, the distribution of alleles in the two parents cannot be known and has to be inferred from the progeny.

This complexity is compounded by the large number of marker loci that may be segregating in any given cross. Fortunately, a range of software packages are available to estimate these recombination frequencies, identify linkage groups, assign the markers to the most likely order and space them in map units (cM) on these linkage groups; examples are MAPMAKER and JOINMAP. Ideally the number of linkage groups should be equal to the chromosome number in the gametes but, unless the markers available provide a good coverage of the genome, it is likely that those markers on a given chromosome may appear as two or more separate linkage groups simply because the subsets are not sufficiently close for them to be recognized as being together on one chromosome.

A typical marker framework map is shown in Figure for the chromosomes of *Brassica oleracea*. Clearly these maps provide a very detailed coverage of the chromosome. Many of the markers are very close, *i.e.* less than 10 cM, and over this distance it matters little which of the two mapping functions is used. With the more widely spaced markers, *i.e.* recombination frequency >15%, Haldane's function will exaggerate their distance apart because it will allow for more double crossovers than actually occur.

As more markers are mapped, the total length should converge on the value predicted by the chiasma frequency, *i.e.* approximately 2 3 n 3 50 cM. It is important to remember that genetic map distances have standard errors which are dependent on the size and type of population used to construct the map, the mapping population. The map obtained is that which best fits, given the data and the assumptions underlying it, but there may well be other maps which also fit the data well, though are slightly less likely. Any particular map always develops a greater aura of respectability once it is published, frequently without stating its reliability, and subsequent yet different maps are treated with suspicion. A recombination frequency of true value p has a standard error of ?[p(1 2 p)/N], where N is the size of the gamete population. For example, if p is 0.1 (*i.e.* RF = 10%) and$N$ = 100, the standard error is 0.03 or 3%. In other words the 95% confidence interval of the estimate lies approximately between 4% and 16%.

There are other reasons, apart from statistical sampling, why maps could be wrong. The most obvious is the accuracy with which the data are scored and recorded. Autoradiographs and banding patterns on gels can easily be misread and research workers are loath to discard data even though its interpretation is ambiguous. All data should be checked independently by two or more people and any ambiguities removed.

Wrong scores will bias the data and often exaggerate map lengths as they suggest double recombination; wherever double recombination appears to have occurred in two short, adjacent intervals, the data should be checked. Changes in methylation rather than scoring errors could cause a restriction site to appear or disappear so giving the false impression of double crossing-over, or it could simply be an error.

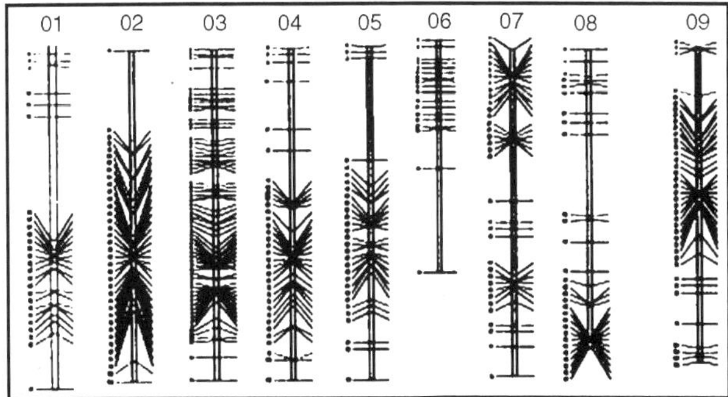

Fig. Typical Molecular Marker Framework Map.The Nine Groups of Brassica Oleracea Based on RFLP, Isozyme and Morphological Markers

Where maps appear to have major inconsistencies in different crosses, different chromosomal structural arrangements may be responsible, such as translocations or inversions. There is considerable evidence that chiasma frequencies, and hence recombination frequencies, may be quite different in male and female meioses even on the same plant although in other studies this may not be so. Thus although the order of markers on a linkage group may remain the same, the relative distances between them may be quite different if the map derives from male versus female meioses. For example, in *Brassica* it appears that the genetic map obtained from a backcross where the F1 is the female parent is 60% longer than when the F1 is the male parent. It is also clear that environmental factors during meiosis, particularly temperature, can affect the number and distribution of chiasmata, and this could be very critical if the crosses are set up at different times of the year. Two or more different populations will almost certainly be segregating for different combinations of polymorphic markers, although some at least should be in common.

These common ones can be used to overlay the two or more maps and a consensus map produced from an amalgamation of these. Again software is available to produce these consensus maps. Despite the various error-causing factors discussed above a considerable degree of consensus can be produced from such populations and, considerable similarity of chromosome sequence is conserved between even distantly related species, allowing the potential for cross-species genetic transfer.

It was stated above that the total map length should be dictated by the chiasma frequency, in so far as this can be accurately measured. When the first few genes were initially put onto genetic maps in the first 40 years of this century, they only covered a small part of the genome – except in wellstudied species such as *Drosophila* and maize. As more and more genes were placed on the map, so the total map lengths increased towards the asymptote dictated by the chiasma frequency. However, during the first decade following the use of molecular markers, the map lengths of some species appeared to exceed that expected, and it was even argued by some that the chiasma theory of recombination might be wrong.

No one would wish to argue that chiasma frequency data are free from error but there would have had to have been quite excessive underscoring to result in the disparity which was occasionally found. It would appear, however, that the excessive lengths were due in part to errors in scoring and to the use of small mapping populations, because subsequent map sizes have shown a progressive decrease towards the predicted asymptote. Scoring errors bias estimates of RF generally upwards, while RFs have to be large to be detectable in small populations. Our own experience with mapping in cereals and brassicas has also shown a progressive decrease in estimated map length as more markers can be found to fill some of the large gaps in the map and as errors are removed from the mapping data.

CHARACTERISTICS OF THE GENES

When genetics was a young science most of us considered the genes as the determinants for the "unit-characters", each gene responsible for a separate quality of an organism. The facts of Mendelian inheritance showed us that there are very small hereditary differences—differences between a contrasting pair of characters.

And it seemed obvious to assume that in such a case the genes themselves occur in pairs, one member of the pair causing white flower-colour and one blue colour, one causing waxy endosperm in corn and one horny. It is very curious that this idea should have taken such a firm grip upon the imagination of the earliest geneticists. If we compare genetics with other sciences—with physics, chemistry, astronomy—we generally see that it is not self-evident to assume that where we see a difference, or where we see two contrasting qualities, two different contrasting causes occur.

We speak of hot and cold, of light and dark, of wet and dry. But we do not really think of a separate cause for wetness and a separate one for dryness—a relative presence or absence of water does very well as an explanation—and the same is true in all the other examples. To assume the existence of pairs of genes in each pair of which the resultant action would be contrasting is an unnecessary complication. The alternative is a much simpler hypothesis which was proposed by Bateson and by myself, and which went under the name of the "presence- and-absence "theory of the gene. We assumed that in a monogenetic difference—the smallest "Mendelizing" difference—we were dealing with the absence or presence of just one gene. At first this hypothesis met with a veritable storm of protests.

To some of my colleagues it seemed the height of folly that the absence of anything should ever determine an inheritable quality. On the face of it, this objection seems very strange; for it seems quite logical to speak of silence as determined by a purely negative absence of noise, of dryness determined by the absence of water. But of course the apparent absurdity of thinking of the absence of a gene as causing an inheritable quality grew out of Weismann's conception of determinants, inherited protoplasmic entities that should be each responsible for just one particular property of a living being.

The idea, now generally accepted by biologists, that the qualities of an organism develop as a result of growth and development, this development being caused by a co-operation of thousands of factors, some of these factors being environmental, while some others—the genes— influence the course of development from within. If we compare a plant having pale-pink flowers with one having dark-pink ones, we assume that the development which leads to the production of the plant with the pale flowers is due to the co-operation of thousands of causes and substances. But in the other plant one additional substance is present, whose action at the appropriate developmental stage allows a further transformation of the pink colour into a darker one. It is therefore not the absence of this gene which directly causes pink colour; it is the presence of the thousand-and-one genes which both plants have in common.

If we knock off a man's hat we may see that he is red-headed. He does not become red-headed as a result of the absence of the hat, but we can see the result of all the other things which cause his hair to be red, provided we take away that hat. If in the course of this story we substitute an inherited gene which helps to transform red hair into black, the reasoning remains essentially the same. Very soon after the original presence-absence hypothesis was first published it was found that in its original shape this hypothesis was not adequate to explain all the facts. It was found that very often in inheritance we are not dealing with just two contrasting qualities, but with three or more alternatives. In snapdragons we have a series of varieties that gradually grade into each other, starting from yellow through different intermediate stages to red.

If we call those forms A, B, C, we find that any pair of them between them show a difference in just one gene. When we mate A with B, we will get only A and B plants in following generations; if we combine B with Z, no 's or C"s, but only B's and Z>'s will be found in the second generation. The best explanation of such series of multiple alleles is this: that in such cases we are dealing not with the presence and complete absence of one gene, but with different quantities of one and the same gene in the one spot on the chromosome where our gene is located. Instead of the presence-absence hypothesis, we now prefer to speak of a hypothesis about quantitative differences in inheritance.

It is highly probable that in certain instances we are really concerned with presence versus complete absence; in other cases, however, intermediate stages may occur in which fewer molecules of the gene are present per cell-nucleus. This quantitative gene-theory is closely bound up with the hypothesis and which is now generally accepted by geneticists—namely, that the genes are not living protoplasma globules, but that they are non-living chemical substances, that have only this in common: that they are each the ferment for their own synthesis. With Alexander, we may speak of reproductive catalysis. We assume that each molecule of a gene, between every time the cells, in growing tissue, divide, attracts and helps to combine from the ambient building stones just one new molecule that will in every particular be its exact duplicate.

The genes are combined longitudinally in long series—the chromosomes—and it is the chromosomes whose movements and combinations give rise to the phenomena we notice as Mendelian segregation and redistribution of genes. The number of the chromosomes is highly constant for every plant species; and sometimes, as in wheat and barley, different sections of the group are distinguished by different numbers of chromosomes. In that case the numbers are always multiples

Many genes of different kinds of a common basis number, and it is very probable that in such a series the higher chromosome numbers may have originated by reduplication of entire chromosome sets during the course of evolution of the group. In the cells of the plants those chromosomes, made up of genes, occur in pairs. The homologous chromosomes of each pair are not necessarily identical. We know that in a plant heterozygous in respect to one gene, the chromosomes of the corresponding pair are different in that particular spot ("locus "). The two chromosomes of a homologous pair have originally been derived from the two parents, one from the pollen-cell and one from the ovule.

The same is true of all the other pairs. The somatic (body) cells have a double set of chromosomes; the germ-cells have only a single set—one of each kind. In ordinary growing tissue every chromosome doubles its mass (and probably the quantity of every gene present on it) between two cell divisions. The two new chromosomes each consisting of part of the two original ones are identical, the break occurring on each at exactly the same spot. And when cell-division commences, the chromosomes are drawn apart to different poles of the cell in such a way that each daughter cell obtains a complete set of chromosomes that is identical with the original set.

The exception is that a "reduction division" occurs when germ-cells are formed. In this case there is a second cell division without a preceding chromosomal doubling. At this reduction division the two chromosomes of a homologous pair become separated one from the other. The result is that in a plant, heterozygous in respect to one gene, Ay in which all the somatic cells are Aa, the germ-cells will be either A or a or only half the germ-cells contain the full quantity of gene A.

When the break does not happen to occur at exactly the same "locus" in the two original chromosomes, one of the new chromosomes may have one gene in duplicate and the other one may be lacking it. Mutations may arise in this way from small irregularities during crossing-over. for this reason, be of four kinds: ABy Ab, aB and ab.

To a certain extent this redistribution of distinct genes over different germ-cells results in a redistribution of the qualities which were shown by the parents of the heterozygous plant. If we cross a plant with AB to an ab plant, the hybrid will be AaBb and as this hybrid will be making four kinds of germ-cells, there will be all sorts of offspring, with different combinations of the qualities in which the grandparents differed.

We were just speaking of genes located in different chromosomal pairs. But sometimes plants are heterozygous in respect to two genes that happen to be located in the same chromosome. In this case those chromosomes tend to be transmitted as such, in their entirety, so that the distribution of the two genes is not mutually independent. Such genes are "coupled ".

All kinds, all degrees of coupling have been met with, from complete or almost complete coupling to very loose coupling. Morgan proposed a hypothesis which is now very well proven and established—namely, that the degree of coupling is dependent on the greater or lesser degree of contiguity of the genes. Genes that lie close together are closely coupled, those that are far apart must lie farther apart on the same chromosome. The mechanism which is thought to account for coupling is the phenomenon of "crossing-over ". Sometimes chromosomal strands seem to touch at one or more places, and when the strands break apart, the top half of one strand may be attached to the bottom part of the other. A new pair of chromosomal strands originates, but if on both sides of a break there are loci where the strands differ in chromosomal content, those loci are redistributed, so that, as far as the independent inheritance of separate genes is concerned, crossing-over may help to redistribute the genes, even if they are located in the same chromosome.

GENETIC VARIATION

Our main interest is to isolate genes that are responsible for the intra- and interspecific morphological variation in plants. Therefore, the analysis of natural genetic variation is the starting point for the majority of individual projects in our lab. For the isolation of genes involved in morphological variation, exploiting natural genetic variation offers some distinct advantages as compared to classical mutagenesis experiments. Most importantly, it allows the identification of naturally occurring, differentially active gene alleles that are likely targets for the evolution of morphological variation. Because naturally occurring lines do not carry major deleterious mutations that impair their survival in the wild, the approach also counter-selects against the isolation of mutant alleles that result in strong phenotypes as commonly detected in mutagenesis experiments.

This applies in particular to wild populations with a high level of inbreeding, such as in our favourite model organism, *Arabidopsis thaliana*. So why has this approach not been followed earlier? Mainly because the isolation of genes that modify quantitative traits has technically not been feasible. However, the rich biological and genomic resources, such as the fully sequenced genome, have turned Arabidopsis into an ideal model system to investigate the molecular basis of natural genetic variation at the gene level.

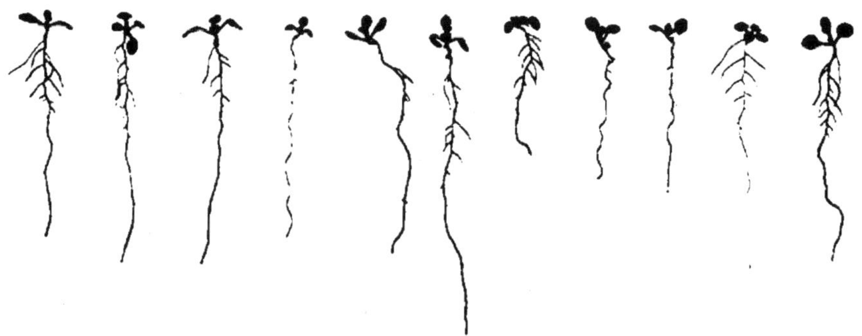

Fig. Natural morphological variation of root system architecture between wild isolates of Arabidopsis. Most of this variation results from the influence of genetic factors.

Plant organs are formed in a continuous, post-embryonic manner by ordered cell divisions and expansions. The extent of growth ultimately determines organ shape. Thus, we are primarily interested in genes that modulate growth rate. To isolate such genes, we focus on the root system. This is because of our interest in root biology per se, but also because roots grow from a so-called meristematic region at their tip, which produces concentric tissue layers through controlled cell proliferation, elongation and differentiation. This feature enables us to reduce the problem of accurately measuring growth from three dimensions to a single one.

Nevertheless, analysis of root system growth requires care and highly reproducible growth conditions, because root system development is very plastic and responds to many macro- and micro-environmental cues. The natural variation in root system morphology between Arabidopsis accessions collected from different locations and grown in identical growth chamber conditions on tissue culture media.

So far, we have isolated an important regulator of root system architecture from the accession Uk-1. This gene is a major so-called quantitative trait locus (QTL), which we gave the name *BREVIS RADIX* (*BRX*), latin for "short root". Other efforts in this area include attempts to isolate two other QTLs for root growth located in other accessions.

ACTION OF PLANT HORMONES

Plant hormones are at the heart of many plant developmental processes. They determine critical growth parameters, such as the rate of cell division or elongation, as well as developmental decisions, for instance where along a primary root a lateral root should be formed. Different plant hormones control different developmental processes to varying degrees, sometimes in the same, sometimes in opposite directions. Whether and how the overlapping influence of different hormones on traits is coordinated is largely unknown.

Part of the research in our lab aims to determine how the major plant hormone auxin, arguably the most important in shaping the plant, regulates downstream events that lead to discrete morphological change. After all, those changes require structural rearrangements on the cellular and organ level after hormone signaling. One tool to answer this question are the *hy5* and *hyh* mutants of Arabidopsis, which display constitutively increased auxin signaling.

Another important question is the nature of crosstalk between hormone pathways, that is the impact of more than one hormone signaling pathway on the same trait or molecular event. Does such crosstalk simply reflect a quantitative add up of different hormone signals at the point of convergence (In the way the water flow rate through a river is determined by the contribution of its two tributaries downstream of their junction), or is there real crosstalk happening upstream (The two tributaries that will meet have built-in dams, and if so, do the dam gatekeepers communicate with each other to intentionally regulate the water flow beyond the river junction)? Much to our surprise, we isolated an important regulator of hormone pathway interaction in the root through our natural variation project: loss of function of the *BRX* gene results in reduced root growth because *BRX* is required for the biosynthesis of the hormone brassinolide, whose concentration is in turn rate-limiting for auxin action.

Since *BRX* is itself under auxin control, this creates a feedback loop required for optimal root growth. Whether an equivalent loop works in the shoot, and whether this feedback is conserved in other species are some of the questions we are currently interested in with regard to this project, generously funded by the Swiss National Science Foundation.

Variation

If you have grown Fast Plants you will have noticed variation among the individual plants within the group. Variation can range from a little to a lot. Variation is one of the fundamental characteristics of life. All organisms exhibit some variation among individuals. Understanding the ways that variation is manifested in organisms, how it comes to be expressed through the development of the individual, and how it is transmitted from one individual to the next generation of individuals are central themes in biology.

Working with Fast Plants will enrich a student's understanding of variation. By observing the growth and development of a Fast Plants through the various stages in the life cycle, students will become aware of many visible features, or phenotypes, that make up the organism. Only, however, upon close observation of a population of plants, will they become aware that the characteristics observed on one plant vary more or less on other plants. Such is the nature of variation.

PHENOTYPIC VARIATION

Phenotypic variation, e.g. plant height at a particular stage of development, is considered to be the expression of the genetic makeup (genotype) of the individual as it interacts with the environment. Variation in plant height among individuals in a population is therefore due to variation in the interaction between the genotype and the environment, as expressed through the development of each individual plant.

In order to be useful in an experiment the phenotype must be described using terms that are widely understood and easily communicated. For these reasons scientists have agreed upon various standards or descriptors to describe characteristics in the natural world. The choice of how to describe what you observe is important, because it will determine the kinds of descriptors used and establish the basis for recording, analyzing and communicating results.

ENVIRONMENTAL VARIATION

Much can be learned about the role of light, temperature, and nutrients on plant development from experiments in which one or more environmental parameters are changed. Because Fast Plants are highly responsive to changes in the environment, they are ideal for examining the role of the environment on the expression of phenotypic variation.

Although Fast Plants are able to grow within a wide range of environmental conditions, for most investigations it is recommended that they be grown under uniform and ideal conditions. In this way variation arising from sub-optimal conditions of environment will be minimized.

The Wisconsin Fast Plants Information Document (WFPID) Understanding the Environment describes how to provide and maintain the various physical, chemical and biotic components of the environment that are most suitable for Fast Plants.

GENOTYPIC VARIATION

How can students use Fast Plants to investigate the contribution of the genotype to the phenotype? Fast Plants, rapid cycling Brassica rapa, are genotypically variable in that they have a genetically controlled mating system that prevents self-fertilization and favours out-crossing among individuals. As a consequence, even seed stocks selected for uniformity of specific phenotypes and genotypes exhibit considerable variation for other traits.

Fig. Rosette, dwarf, basic, elongated internode

The Wisconsin Fast Plants® Programme has developed a number of genetic stocks of rapid cycling Brassica rapa for genetic investigations on the nature and inheritance of variation. Some WFP stocks contain distinctive mutant phenotypes, e.g., anthocyaninless plant, anl, yellow green plant, ygr1, rosette, ros, and male sterile, mst2, that are suitable for Mendelian genetics. Other stocks exhibit phenotypes whose expression may vary continually and which may be quantified as discrete or countable units, e.g., number of hairs, or as estimates of size, e.g. petite dwarf, dwf1, or of intensity of colour saturation, e.g., purple anthocyanin. These quantitative phenotypes may be conditioned by a few or many genes and normally require numerical descriptions in which the statistical notations of population size, (n), range (r), arithmetic mean (x), and standard deviation (s) are applied. Yet other stocks have been developed to combine both simply inherited mutant genotypes and quantitatively expressed phenotypes. An important part of WFP is the continuing development and improvement of seed stocks for uses in genetics.

11

Tissue Modifications Caused by Parastites

The tissue-changes that may result from parasitic attack on plants, remarkable though they may sometimes be, resemble in many respects those that have been described above, allowance being made for the fact that the stimulus may be more localised and may continue for a longer time than that caused, for instance, by wounding. It seems very doubtful whether there is any essential difference between the modifications of tissues and organs found in galls due to parasites and those that may occur in wound calluses or the various other types of reaction to wound injury or physiological causes.

Hypoplastic modifications are seen in the sterility caused by some parasites, in the small size.of the leaves in trees attacked by certain virus and other diseases and in inhibition of tissue differentiation as a result of parasitic attack. An interesting example of the latter is the prevention of the formation of an absciss layer in cherry leaves attacked by *Gnomonia erythrostoma*, with the result that they do not fall in the autumn but remain hanging on the tree and serve to infect the new leaves in the following spring.

Host-tissue reactions to parasitic attack by the formation of wound gum in woody tissues and wound cork in all living parts resemble those due to simple wounding in their main characteristics. The band of reactionary cork formed around a leaf infection can sometimes be seen in thin leaves as a narrow dark circle on holding them up to the light. The cork may be effective in checking the parasite or may fail. A common result of its formation in leaves is the necrosis and dropping out of the surrounded area, producing what is known as a 'shothole' effect.

It has been already mentioned that there is a tendency for the tissues of reaction to non-parasitic stimuli to consist at first of elements resembling one another, as, for instance, in calluses and intumescences, in the repaired epidermis, palisade and spongy parenchyma at the edge of a leaf wound, or in the thin-walled derivates of the cambium on the xylem side in the thickened shoots of kohlrabi prevented from flowering in the second year. This tendency is still more marked in many galls and has led to the use of the term' gall' or 'tumour' tissue to characterize the masses of soft rounded cells, often with few or no intercellular spaces, that may develop as a result of the presence of a parasite. Galls are swellings on plant organs due to a stimulating or irritant action of parasites such as fungi, bacteria or insects on the tissues. In a few cases they are caused by viruses.

They may vary from tiny swellings involving a few cells of the epidermis as in the early stages of wart disease of potato to the soft rounded tumours several inches across of maize smut, or the large woody masses formed sometimes in crown gall or in certain rusts on Acacia. The term gall is sometimes extended to include organic outgrowths or proliferations, such as the leafy organs formed in cereal ears attacked by *Sclerospora* and also to cover cases of uniform enlargement of an organ, as in flowers of some *Cruciferae* infected by *Cystopus candidus,* or the 'bladder plums' produced by *Taphrina pruni.*

Galls due to Tissue Swellings: *'Histoid' Galls*

Some of the simplest galls are caused by species of Synchytrium. The fungus may be confined to a single epidermal cell which generally becomes much enlarged. Neighbouring cells also react, usually hyperplastically, and the host cell becomes surrounded by a sheath which, as in *S. taraxaci* on the dandelion, may be derived from both epidermal and subepidermal cells.

The bordering cells in this little gall elongate tangentially to the host cell and divide by walls at right angles to the latter. A small-celled sheath of two or three layers in depth is thus formed around the enlarged cell containing the fungus. No intercellular spaces occur in the sheath nor is there any detectable difference between the derivatives of the epidermis and mesophyll that constitute it; there appears to be also no tendency to form cork, perhaps because of the short duration of active stimulation.

Another very simple epidermal gall is caused on poplar leaves by *Taphrina populina,* syn. *T. aurea,* the mycelium of which is situated between the cuticle and epidermis and for a short distance down between the cells of the latter. These cells elongate to two or three times the normal length and may divide once or twice, the result of their uneven growth in the developing leaf being the formation of a small bulge on one side of the leaf and a corresponding depression on the other.

Another leaf gall of a simple type is that formed on apple leaves by the aecidial stage of *Gymnosporangium juniperi-virginianae* in America. The affected part of the leaf is swollen to two or three times its normal thickness by a great increase in height of the spongy mesophyll cells, the palisade being little altered. The elongated cells divide by transverse or oblique walls and the intercellular spaces are lost. In the thickened leaves produced on the shepherd's purse by infection with Cystopus candidus, the mesophyll is composed wholly of large round cells, distinction between palisade and spongy cells being lost.

Taphrina deformans also thickens the peach leaves that it attacks, and causes the palisade cells to multiply and lose their normal shape, becoming almost isodiametric in the outer layer, where they are hard to distinguish from epidermal cells. *Exobasidium oxycocci* makes all the mesophyll into a uniform type of parenchyma without intercellular spaces in *Vaccinium macrocarpon,* one of the cranberries, but in the thickening caused by *E. vaccinii in V. vitisidaea* leaves, the palisade is little altered and the swelling is due to the development of roundish or polygonal cells with no intercellular spaces in the spongy parenchyma. A simple stem gall is the common one on goutweed or bishopweed caused by *Protomyces macrosporus,* where there is a hyperplasy of the outer cortical cells, restricted to those in the neighbourhood of the intercellular hyphae. The easiest way to locate the hyphae of this fungus in the early stages of infection is to look for a fully formed cortical cell in process of division.

Somewhat more complex are the small corky swellings that constitute the disease known as scab on the leaves and twigs of citrus plants. The parasite, a conidial stage of the fungus *Elsinoe fawcetti,* causes in attacks on the under surface of the leaf a necrosis of a few superficial cells, and provokes a hyperplasy of the underlying spongy parenchyma. The new cell walls are roughly parallel to the necrosed surface, and the reaction may extend through the leaf and cause some of the palisade cells to divide. Cells near the lesion elongate considerably and divide by several cross walls. A band of the new cells thus formed becomes a cork cambium. This extends up to the epidermis all round the lesion, and cuts off cork on the side next the latter and a little phelloderm towards the sound tissues. All the primary cell walls of the hyperplastic area are thickened, and intercellular spaces are much reduced.

A good example of a gall restricted to the vascular region is that produced by *Sorosphaera veronicae* on *Veronica chamaedrys,* the germander speedwell. The swelling starts in the procambial region of the shoot, from the point behind the apical meristem where the xylem is just beginning to form, and extends backwards without involving the cortex or pith, and sometimes outwards along a young leaf bundle. Spiral vessels can be found throughout its length but do not occupy any fixed position in relation to the mass of gall tissue, being sometimes on its inside towards the pith, sometimes on its outside. Towards the cortex, sometimes in the middle. This shows that the gall tissue is formed not only from the cambium, but from cambial derivatives lying around the protoxylem vessels. I t consists of more or less polygonal cells of a very uniform type, the only other elements found in the swollen part being the few xylem elements that were already differentiated before the parasite reached the area, and some scattered groups of elongated cells. The parasite, an intracellular one, does not appear to occur in all the tumour cells, but the hypertrophy and hyperplasy are general throughout the region involved. The cortex and pith usually undergo no histological modifications, though there are indications that, in later stages, the parasite may reach the inner cortex by passage from within through the endodermallayer. In the finger-and-toe or club-root disease of turnips, cabbage and other crucifers, the parasite Plasmodiophora brassicae is again mainly confined to the central cylinder. In inoculations on cabbage stems the fungus reaches the cambium fairly soon and induces a marked hyperplasy in the cells in this region. The elements cut off on the xylem side remain undifferentiated and again rapidly divide. In older stems the medullary rays are chiefly infected and show great hypertrophy and hyperplasy. Except at the point of original invasion, the phloem and secondary cortex only become infected outwards from the cambium.

In roots the bulk of the hyperplasy is in the cambium and the phloem initials adjoining it; the new cells in the position of the phloem are thin-walled and accumulate starch. The increase in size of the medullary rays may divide the vascular ring into widely separated bundles which ultimately may become much distorted by the uncontrolled growth in the rays. New tracheidal tissue develops in the rays. The changes that occur in the vascular region in this disease show a certain similarity to those already mentioned as occurring in allied plants as a result of physiological processes consequent on the prevention of flowering. In various other types of gall, insect as well as fungal, the tendency already mentioned for the cambium, when stimulated to abnormal activity, to cut off non-lignified parenchyma without fibres and with fewer and thinner-walled tracheids on the xylem side, and secondary parenchymatous phloem outside is strongly marked.

In some of those that cause the medullary rays to enlarge, so as to separate the vascular ring into isolated bundles, these tend to become bordered by a cambium forming in the ray cells and in the region of the protoxylem, very much as was found in the thickening due to prevention of flowering in kohlrabi. The underlying causes of these processes are not known, but the similarities are sufficient to suggest that whether the stimulus is a fungus or an insect or purely, physiological', the response follows along certain broadly defined lines.

In some fungal and bacterial galls all or nearly all the tissues may become involved in the swelling. The thickened axial shoots that are the most prominent feature of the destructive witches' broom disease of cacao in the West Indies and South America, due to *Marasmius perniciosus*, are good examples of this. As fuany as a thousand of these may be found on a tree. The swelling may involve every tissue from epidermis to pith, sometimes almost equally, each being two or three times the normal thickness, sometimes mainly in the cortex or the xylem. Most of the thickening in the xylem seems to be from cambial activity, but in the other tissues it is mainly due to division of parenchyma in situ. The pericyclic region and deeper tissues may proliferate to such a degree that the cortex is crushed, or may become so parenchymatous that it is almost impossible to define the inner limits of the cortex, though scattered groups of pericyclic fibres can still be distinguished. All the parenchyma cells, except those nearest the epidermis, assume the same shape, elongated radially and with thin walls. The xylem vessels are broader and thinner than usual and the part of the xylem near the cambium is largely parenchymatous.

The fibres become at first septate, then reduced in numbers, and finally are not formed. As the thickened shoot grows in length, tissue differentiation is decreased. In the swollen leaves the distinction between palisade and spongy parenchyma is lost. The 'brooms' are short lived and remain soft and fleshy, cork being rarely found in them. The huge *Uromycladium* galls formed on species of Acacia in Australasia start usually by a proliferation in the phloem, infection beginning generally before cork has developed. The outer tissues are deformed and ruptured by pressure from within, due both to expansion of the phloem and pericycle and to the production of wound wood composed of unlignified rows of tracheids, with a few wide vessels and an excess of parenchyma.

The identity of the original cambium is lost at an early stage, and 13;ter it becomes impossible to distinguish clearly between phloem, cambium and outer xylem. Cork may form in the pericyclic region or the outer cells may continue to proliferate as hyperhydric tissue. Tracheids develop in any of the tissues that take part in the formation of the gall. No plant gall has been so completely studied as crown gall caused by *Bacterium tumefaciens,* because of its intrinsic interest, its wide range of host plants, some of considerable economic importance, and because of suggested analogies between it and cancerous growths in man and animals.

The organism which causes crown gall is one which is not ordinarily provided with enzymes capable of dissolving the polysaccharides of cell walls or starch, so that it does not form cavities or penetrate into the interior of the plant cells or induce soft rotting. On the other hand, it has marked powers of stimulating the cells in its immediate vicinity to divide and form hyperplastic masses of tumour tissue. When it is inoculated into the tissues of herbaceous stems that are still growing, it may occur in the intercellular spaces of considerable lengths of stem formed subsequently to the inoculation and it may also spread actively or passively in the fully grown internodes, along the xylem or where, as in the pith or cortex, intercellular spaces are in erotical communi-cation with one another for some distance longitudinally.

In these cases it may give rise to continuous strands of tumour tissue formed by the division of bordering cells, or it may only cause visible hyperplasy here and there along the course of its extension. It is not uncommon to find, following inoculations of this type towards the top of the stem, closed vascular rings, spheres or cylinders in the pith and cortex. When these are found in the pith the orientation of the tissues is frequently inverted, the phloem being inside and the xylem outside the cambium. In the cortex and petiole the orientation is normal, and in the latter situation the stimulation from the parasite may lead to the formation of a closed concentric bundle by division of the ground cells of the petiole outside the leaf trace, just as occurs when a petiole is rooted as described above. The centre of the new formation is a mass of tumour tissue in which cells of the ground parenchyma not transformed into tumour cells may sometimes be seen. In some of these infected petioles the organism has been found to occur in the protoxylem, and the meristematic divisions in the ground tissue begin opposite this point.

Tumour strands in the outer cortex resulting from inoculations of *Bact. tumefaciens* in this region have been found to arise from extension of the bacteria along intercellular spaces running longitudinally in the stem or drawn out by "rapid elongation of young parts, the cells bordering the infected spaces being stimulated to active division. A secondary gall has been found to develop at some distance from the point of inoculation, and in this a closed vascular sphere may occur, having no vascular connection with the central cylinder of the stem. In some pith strands the cells bordering the first-formed tumour cells may show only hypertrophy but in others a vascular cylinder develops, with xylem outside.

The vascular tissue may form around a core of tumour cells in a part only of these strands; in strands in the pith and cortex it develops independently of the original cambium and usually at some distance from it. The inverse arrangement of the bundles thus formed in the pith resembles that found normally in many families of plants possessing medullary bundles. Sometimes also, both in normal pith and in the protoxylem of inoculated plants, only a little nest of phloem occurs without cambium or xylem, and it is interesting to note that a similar formation has been found in the middle part of the protoxylem parenchyma in kohlrabi plants prevented from flowering.

The exciting cause of these inverted normal or induced formations appears to be unknown, though attempts have been made to explain them. When crown gall results from inoculation into the deeper layers of a herbaceous stem the tumour tissue may appear to burst through the central cylinder and cortex to the outside. This appearance is due to a progressive hyperplasy of the ray and cortex cells, the outer layer of the latter being sometimes crushed by pressure from below. In such erupting galls it is easy to follow the conversion of the normal cells into gall tissue by multiple division. In the tomato, the gall cells thus formed become smaller and smaller, by division, for some weeks, then enlarge somewhat; an average normal cell of the cortex may be nearly 5000 sq. μ in area, and this may fall by division to 166 μ after about a month and rise to over 400 μ in about two months. The tumour tissue from secondary tumours in herbaceous plants appears to be able to continue growth after removal from the infected plant. It still goes on growing when, after several' generations', bacteria cannot be isolated from the mass of tumour cells and the ground-up mass is incapable of infecting healthy plants. The apparently sterile tumour mass can be implanted by grafting into a healthy sunflower and will continue to grow into a tumour resembling a typical crown gall on the plant, reaching sometimes a diametre of I cm. after seven weeks. Work at East Mailing has shown that similar tumours can be produced by the growth-promoting substance indolebutyric acid, while in America it was found that indoleacetic acid causes both primary and secondary tumours, the latter at some distance from the treated site.

Though it has been found that *B. tumefaciens* can form indoleacetic acid in media containing tryptophane or other nitrogen compounds of high molecular weight, and the accumulated evidence indicates that a diffusible substance is concerned in the development of the tumours, it is not yet clear what the substance is or whether it is a product of the host cells or of the metabolism of the bacteria, or even if it is one of the recognised growth promoting substances, such as indoleacetic acid, at all.

In woody stems such as those of the raspberry or blackberry, in which crown gall is not uncommon, the galls may appear as multiple outgrowths. In these infections the swelling may at first be confined to the region of the pericycle outsidc the fibre bundles. There may be no nests of tumour tissue, as in.the herbaceous stem galls, the swelling consisting of thin walled radially elongated parenchyma in rows and with scanty contents. The multiplying cells lie below the cork, in the pericycle, which form in layers in the neighbourhood of the endodermal region.

At this stage the new tissue may be regarded as a parenchymatous derivation of the actively dividing pericycle, but the reaction soon extends down between the fibre bundles and involves the tissues between them and the outer part of the phloem. The growth in this position bends up the pericyclic fibres and separates groups of them widely from one another. Here and there, usually not over the whole arc underlying the gall, the cambium is stimulated tq form wound wood, which is sharply differentiated from the normal xylem by a scarcity of fibres, the thinner walls and wider lumina of the tracheids, and the absence of large vessels. In the tumour mass outside the cambium single tracheids or tracheid groups are formed by differentiation in individual tumour cells, just as occurs in an ordinary callus outgrowth and, as in callus, the tracheid islands may become bordered by a meristem producing further tracheids on one side and phloem or gall parenchyma on the other. The gall often becomes highly vascularised in this manner and some of the strands may reach the original central cylinder.

The surface of the gall is usually a loose mass of brown collapsed cells with weakly lignified walls, resembling the hyperhydric tissue frequently formed on the sur face of a callus. There is, indeed, a fairly close histological analogy between crown' gall and wound callus of the type found causing knots on apple grafts. There is little reason for the supposed analogies between this plant gall and the malignant growths that form animal cancers. As will have been seen, the processes involved in its formation are not essentially different from those found in other plant galls. Similarly, the leafy outgrowths that sometimes occur on the galls, and that have been termed embryomata from a suggested resemblance to one type of cancer, are no doubt to be likened to callus shoots.

Organoid Galls

There is another type of structural modification sometimes produced in plants as a result of the presence of a parasite in the tissues to which, though the changes are morphological rather than histological, the term' gall' is often applied. These, organoid' galls, as distinct from the' histoid' type already discussed, result especially from systemic in actions with obligate parasites of the kind that do not kill the invaded plant parts at an early stage of parasitism. They are characterized by changes in the morphological plan of the shoots or other organs affected, or in the rhythm of their development. Marked changes in symmetry and habit may be found in plants affected by some rusts and mildews.

A good example is the change from the normal short unbranched stem, with radical leaves, of *Launea asplenifolia,* a common Indian weed, to an elongated much branched axis, with cauline leaves, when infected with *Puccinia butleri,* a systemic rust. Stem elongation is caused by *Ustilago hypodytis* on *Elymus arenarius* and *Sclerospora sacchari.* on sugar cane and the number of internodes and leaves is increased by various rusts, e.g. *Uromyces pisi* on *Euphorbia cyparissias* and by *Gibberella fujikuroi* in the 'bakanae' disease of rice in Japan, where it has been traced to the production of a crystalline active principle gibberellin; shortening of internodes occurs in rye attacked by *Urocystis occulta.*

The symmetry of the leaves is altered in some diseases. They may be changed from simple to irregularly lobed in Berberis buxifolia attacked by Aecidium jacobsthalii in South America. In the' reversion' disease of black currants one of the most useful aids to diagnosis is the change in the venation and, especially, the lobation of the leaves of affected plants. The reverted leaf may have only three main veins, diverging from the top of the petiole to the extremity of three main leaf lobes, instead of the usual five of each.

Lateral sub-veins from the middle main vein are reduced tq about half the normal number, but they serve to supply nearly all the teeth on the leaf margin, whereas the 10 to 16 sub-veins of the normal leaf do not supply more than about half the teeth of the apical lobe. The reverted leaf, best seen on wall grown shoots in May and June, is smaller, narrower, flatter at the base, coarser and of a deeper green than normal. An extreme example of the opposite tendency is to be found, as the name implies, in the 'fern leaf' type of tomato mosaic disease while some forms of tobacco virus infection can cause a twisting of the leaf so that the under surface faces upwards. Another remarkable effect of virus infection on leaves is the formation of 'enations "green cup-shaped or blade-shaped protusions, generally on the under side of the leaf. The very remarkable leafy outgrowths from the leaves of bracken caused by the fungus Taphrina laurencia in Ceylon are probably due to processes similar to those causing enations.

Perhaps the most profound modifications of the type under discussion are caused in inflorescences and single flowers by the action of certain fungi and viruses. One of the commonest of these changes is the inducement of 'phyllody' or the transformation of flower parts into leafy organs. This is found in several virus diseases of tomato, sandalwood and other plants, in the flower of the Japanese plum *Prunus mume,* infected by *Caeoma makinoi,* and in certain crucifers attacked by *Cystopus candidus.*

Single flowers may be doubled by the development of additional petals or their symmetry may be changed from the regular to the irregular plan and vice versa; in *Matricaria inodora* attacked by *Peronospora radii* ligulate flowers may be found in the centre of the disc and tubular flowers at the margin). The rudimentary stamens in pistillate flowers of Lychnis dioica may become fully developed when attacked by *Ustilago violacea,* except that the pollen is replaced by spores, while in the contrary direction the maize' tassel' when infected by *U. zeae* may bear female or bisexual flowers, or even normal grain at its base. Some rusts prevent flowering of the host when there is a perennial mycelium extending from the root-stock to the growing point but permit normal flowering when the infection is local.

The effects produced on the inflorescence of certain crucifers by *Cystopus candidus* may be very intense. The flower stalk and axis of the inflorescence may be enormously thickened while the floral parts are wholly or in part fleshy, green or violet and persist instead of falling early. The petals may resemble sepals and the stamens become leaf-like or occasionally like carpels. The latter may be open instead of united into an ovary, while the ovules and pollen grains may be atrophied, thus causing sterility; sometimes the pistil may be much swollen into a conical thickwalled sac containing sterile ovules while the anthers may have supplementary pollen sacs or become like carpels by the development of stigma-like structures at the tip and rudimentary ovules along the margin.

Occasionally the flowers revert to the primitive spiral insertion of their parts. Chlorophyll and starch may be copious even in the petals and stamens; hairs may develop on the anthers, and stomata in the inner epidermis of the enlarged ovary. In this case the tendency to a lessening of differentiation in the tissues of the swollen parts, which is marked, is carried further to a tendency to reduce the differentiation of organs. One of the most striking cases of virescence of the inflorescence accompanied by leafy modifications of the floral parts is the 'green ear' disease of millets in India, Africa, Japan, the United States, and elsewhere, caused by *Sclerospora graminicola.* As the name suggests, the chief symptom of the disease in the commonly affected bullrush or pearl millet, *Pennisetum typhoideum*, is the transformation of the solid spicate ear wholly or in part into a loose green head, composed of a mass of small twisted leaves.

This is due to the replacement of the upper part of the axis of each floret, which normally forms the grain, by a short leafy shoot. The bristles below the glumes are enlarged and contorted and the fertile glumes enlarged and sometimes turned green. The stamens are often converted into minute leaf-like organs with a distinct division between blade and sheath. When not so transformed, they may occur as brown pointed bodies with no trace of the anthers, or if the anther is formed, its size relative to that of the filament may be much altered and pollen may be absent.

The pistil often undergoesedian proliferation, being replaced by a small leafy shoot, but sometimes by a minute branched axis with undeveloped buds, or by a simple, hollow, hornlike growth, which may be a leaf united along its edges. Sometimes two spikelets arise from the pedicel instead of the usual one, and the number of florets on a spikelet in infected ears may be increased from two to three or four. On the other hand, oats infected by *Sclerospora macrospora* in Mississipi bore only a single seed-like structure in the glumes instead of the normal two close together. The median floral proliferation, which is the main characteristic of the' green ear' cereal diseases, occurs also in certain virus diseases.

A remarkable instance of structural modification is caused by an organism first isolated in the United States from fasciated sweet peas. Later, this organism, *Bacterium fasdans,* was found to be parasitic in a number of garden and other plants in England and to cause leafy and fasciated outgrowths from the buds at the crown of the root-stock and at the stem nodes, when inoculated in these sites. In some plants the leafy shoots from the base are so numerous as to appear like moss.- Bud inoculations on seedlings cause galls at the growing point and in leafaxils, with a complete check of normal growth; in other plants fasciated or cauliflower-like shoots are produced. Proliferation is confined to the bud tissues, inoculations elsewhere on leaves, stems and roots producing no effect. Amongst the 25 genera from which *B. fascians* has been isolated are peas, beans, strawberries, chrysanthemums, petunias, and asparagus.

In strawberries a high proportion of inoculated seedlings of the St.-Jan variety developed leafy and floral abnormalities resembling those caused by cauliflower disease, but the inoculation of 'runner' plants gave inconclusive results. The term witches-' broom is applied by the country people in some parts of Europe to irregular tufts of crowded twigs found in trees and shrubs. They have, as indeed have several of the' organoid' type of galls above-mentioned, some of the characters of 'histoid' galls, for the tissues that take part in their formation are often swollen. Their chief character, however, is the production of short twigs in irregular groups, often all turned upwards and with small leaves.

When due to fungi, rusts and species of Exoascaceae are mainly concerned; the fungus passes into the new shoots produced each year, and often fructifies on them. When spore production is finished, the leaves and sometimes the younger twigs die off. Well-known examples are the brooms produced by Taphrina cerasi on cherries and those on silver fir due to Peridermium elatinum. Mites and insects sometimes cause similar growths, and some brooms on pine and spruce are not caused by parasites, but appear to be due to an innate heritable tendency of the tree.

In galls, whether 'histoid' or 'organoid', with rare exceptions, there is nothing new in the tissues or organs of reaction, but merely a disorganisation or an intensification or inhibition of normal processes of tissue and organ formation and differentiation. The parasite can do no more than call out or suppress powers which the cells and tissues already possess, powers which plants can ordinarily make use of in dealing with such accidents as wounding or in meeting extremes of environmental or nutritional or growth modifying conditions. Indeed, most of the modifications in tissues and organs caused by parasites can be paralleled by those brought about by physiological processes without the intervention of a parasite, or by examples selected from the extensive literature of plant teratology.

In the 'quercina' virus disease of the thorn apple *Datura stramonium* the spines on the capsules are eliminated, the stigmatic surface of the pistil is elongated, the tubular corolla becomes one with separate petals, and the leaf margins are eroded. All these modifications have been recently found in a hereditary mutation of the, same host. In speculating on the causes that may underlie some of these modifications, especially those of the type found in organoid galls, it is impossible to avoid the suggestion that they may be due to dislocation in the production or movement of growth controlling substances, much light on the existence and properties of which has been obtained in recent years. Little knowledge, however, is as yet available as to the part they play in stimulating gall formation though, as already mentioned, they seem to be concerned in the production of crown-gall tumours.

It is worth mentioning that phytohormones are formed in abundance by some fungi and may have a similar effect to those in a crude lentil extract in promoting growth in *Nematospora* and fructification in many fungi. It is not only the meristems that have the ability to respond to stimulation by the development of new tissues or by modifications in the products of their activity: other living tissues can playa part not less important. No doubt meristems respond very readily, and there may be indications that the pericyclic region is easily stimulated into activity in some plants, but the examples given above show that this sensitiveness must not be given undue importance. The generalisation has been made that in all living plant cells there slumbers the potentiality for the development of all histological characters that appertain to the particular species. The study of galls gives point to this remark.

THE MECHANISM OF FEEDING BY PARASITES

Parasites reach their food either by sending their mycelium into the interior of the host plant or by remaining for the most part superficial and sending only suckers or a limited growth of feeding hyphae into the surface tissues. Usually in the internal parasites, e.g. *Rhytisma acerinum,* the feeding hyphae penetrate directly into the cells of the host, the cell sap and more solid contents being taken up by absorption after the latter have been brought into solution by the enzymes of the fungus.

Often, however, the earlier stages of feeding are marked by the dissolution of the cell walls of the host by enzymes such as pectinase and cellulase; these walls can provide many fungi with much of the carbohydrate food that they require. In a few fungi, such as the *Exoascaceae,* the mycelium is wholly intercellular, without any intracellular parts, so that absorption of nutrient solutes must occur through the membranes of the host and parasite cells; this is probably a common occurrence, for even when the cells are entered much of the mycelium may remain intercellular and probably receives its food directly from the surrounding cells.

It was at one time believed that some of the superficial parasites of leaves and twigs could similarly feed by absorption from the epidermal cells across the outer wall of the cell and the cuticle, but in most of these fungi, feeding hyphae or suckers have now been found to enter the cell 'cavities; this occurs, for instance, in the light-coloured thread blights and in the dark *Meliolaceae* or *Asterineae.*

In the ectoparasite of some European pine trees, *Acanthostigma parasiticum,* it is stated that the white mycelium which forms cushion-like layers on the under surface of the needles obtains its food through tiny rod-shaped processes which penetrate through only the cuticle and not through the inner parts of the outer wall of the epidermal cells. As in the subcuticular parasites such as *Taphrina rhomboidalis,* absorption of nutrient solutes must occur through the subcuticular membrane of the needle leaf.

HAUSTORIA

Many intercellular parasites that apparently find difficulty in getting adequate nutrition by absorption of solutes into their hyphae from the cavities of the host cells send into the cells specialised hyphae which act as feeding organs. These are known as 'haustoria'. Haustoria occur in all the main groups of fungi, usually but not exclusively in obligate parasites with an ectoparasitic or an intercellular mycelium. They vary in shape from the knob or saclike type found in Cystopus to simple or lobed or branched or coiled hyphae in many *Erysiphaceae* Phycomycetes and rusts, or coralloid types as in *Peronospora schachtii* on beet. Some of the complex forms almost fill the cells they occupy and can hardly be distinguished from an intracellular mycelium, as in *Peronospora parasitica*. Their shape is not very constant, as it may vary to some extent according to the tissue invaded.

The haustorium usually enters the cell by an exceedingly fine hypha resembling the infection hypha arising from an appressorium by means of which, as will be seen later, many parasites enter the outer membrane of the host. After entry the haustorium often grows towards the nucleus of the host cell but while, at first, it may not enter the' primordial utricle' or peripheral plasmatic membrane of the protoplast, only pushing it in and invaginating it, the utricle may finally be pierced by it. In the early stages the haustorium is bounded by a thin limiting membrane, and this sometimes persists. Frequently, however, various thickenings occur around the point of entry or deeper in.

When developed around the narrow style-like stalk of the haustoria of *Erysiphe graminis*, or many rusts, the thickening is continuous with the inner wall of the cell and is closely similar to the infection pegs found where a parasitic hypha enters a cell. It is sometimes effective in checking penetration. When found around the body of the haustorium it is often later in appearing and may, in old haustoria, be a means of walling off the intruder, or a consequence of excessive drain on the cell contents, as in tubers attacked by potato blight; as injury to the cell from over-virulent action of the fungus increases, so the haustorial cover thickens.

There is little doubt that these thickenings are products of the host cell, and indeed it is sometimes possible to distinguish two layers in the sheath, one being the membrane of the haustorium and the other a covering deposited by the host. The latter seems to be usually cellulose, but there may be some lignification of the older sheaths. Callose is stated to occur sometimes between the two layers or in the exterior one. The haustorium itself may have from one to many nuclei. In some smuts the haustoria are typically binucleate, but those of the downy mildews may sometimes have only one nucleus though others have several, as in the coenocytic hyphae from which they arise.

Most of the haustorial fungi arc obligate parasites, but some, such as the Phytophthoras and *Actinonema*, can be grown in artificial culture. It is probable that many of the intercellular hyphae are not wholly dependent for their nutrition on the haustorial apparatus, but, like other fungi occupying a similar position, can get some nutrition by absorption from the surrounding cells. While, therefore, the haustorial habit is not a strict criterion of status as an obligate parasite, it suggests a high degree of specialisation in some requirement which ordinarily it has not proved possible to supply in culture. It marks increased adaptation to the parasitic life over that shown by the destructive action of parasites such as *Pythium* and *Botrytis,* for one of the characteristic features of many of the haustorial fungi, especially the rusts and smuts, is the absence of any gross indication of injury to the protoplast of the host cell for a period which may sometimes be almost equal to the lifetime of the plant. This aspect of their parasitism will be referred to again in a later chapter.

THE FUNGI ENZYMES

In the preparation of their food for absorption fungi, like animals and man, rely on enzymes or ferments, of which they may possess a formidable array. Among the hydrolytic enzymes detected in Polystictus abietinus are cellulase, pectinase, ligninase, diastase, sucrase, inulase, tanninase, pepsin, trypsin, emulsin, and lipase, while the oxidising enzymes of this fungus include catalase, peroxidase, and laccase. In *Aspergillus oryzae* diastase, invertase, maltase, lactase, protease, rennet, and lecithinase have also been recorded.

Many of these enzymes are excreted into the surrounding medium, sometimes in considerable quantity and capable of utilisation on a commercial scale. Apart from the use of yeasts in brewing, distilling and baking, diastase from fungi is used medicinally and in various industrial processes; acetone can be formed by *Mucor rouxii* and other fungi from sugar-yielding plant products; citric acid is now mainly a fermentation product of *Aspergillus niger,* for which one American firm some years ago kept nine acres of mycelium in constant commission; yeasts, Mucors, Aspergilli, and bacteria are all employed singly or in various combinations in the East and in Africa for the preparation of various 'beers', fermented milks, non-alcoholic drinks, and seasonings; while gallic acid, which is used in dyeing and as a constituent of inks, has been commercially produced by the fermentation of tannins.

The proteolytic enzymes of *Aspergillus oryzae* and its allies are used in the Far East in the preparation of readily digestible substances from the protein-rich soy bean. Not less important are the processes of fermentation carried out during metabolism within the hyphae. During the period of stringency of fats in Germany about 1918 more than a million kilograms of glycerine were being produced monthly from fungi, and in Russia *Endomycopsis vernalis* was used experimentally for the same purpose some years later. Molasses, beer-wort, potatoes, hydrolysed wood pulp, or the like, together with salts of nitrogen and phosphorus, provide the culture medium for the fungus. Some moulds are even more effective in synthesising fat as a reserve nutrient in their mycelium than the yeast-like *Endomycopsis;* Penicillium javanicum can store up to 41.5 per cent of fat in its hyphae when grown on glucose and salts.

Yeasts and their allies are also able to manufacture high-quality proteins, with vitamins, from the same raw materials, and some 20,000 tons of dried yeast, containing about half its weight of protein, were annually produced in Germany during the first great war. Probably the best species for this purpose hitherto tested is Torula utilize. These examples, taken from only a few amongst many, indicate that both by means of the enzymes excreted from the mycelium and by those concerned in metabolic processes within the hyphae fungi can synthesise a remarkably wide range of substances from organic materials. Many have been identified and they represent almost every type of compound known to organic chemistry.

The enzymes are organic catalysts which, unlike the inorganic ones, are destroyed by heating. Hydrolysing enzymes split up disaccharides such as sucrose, maltose, lactose, and the like into two similar or dissimilar molecules of monosaccharides-dextrose, levulose, mannose, galactose etc. Cane sugar, for instance, is readily' inverted' by invertase from fungi giving one molecule of glucose and one of levulose, while maltase gives two molecules of glucose from maltose. The hexose monosaccharides, especially glucose, appear to be readily available to fungi; on condensation they give hexosan polysaccharides.

The pentose monosaccharides, such as xylose and arabinose, condense with other sugars to form pentosans- polysaccharides common in pectic membranes, gums, mucilages, and the like; acted on by the pectinase of fungi, they give galactose and other bodies. On splitting up, the polysaccharides give more than two molecules of monosaccharides. The proteolytic enzymes of fungi, pepsin, trypsin, erepsin, etc., change the less soluble proteins into the more soluble and readily diffusible peptones and amino acids. Tannase forms glucose from tannin and lipase, glycerol and higher fatty acids from fats. Certain moulds can grow in the waste liquor from the manufacture of sulphate wood pulp, decomposing the lignin present as chemically combined lignosulphonates.

The oxidative processes of certain bacteria provide energy enabling them to synthesise carbohydrates when carbon dioxide is their sole source of carbon, although they have no chlorophyll and light is not needed for the purpose. Such are the nitrifying bacteria, Nitrosomonas and Nitrococcus, the sulphur bacteria. *Beggiatoa* and *Thiothrix*, and the iron bacillus Spirophyllum. The oxidase of strains of the common mould *Aspergillus niger*, can oxidise glucose to gluconic acid directly. Similarly some yeasts can directly oxidise maltose, giving off carbon dioxide. Yeasts will grow with the organic salt ammonium tartrate as their only source of carbon. Such examples are sufficient to illustrate the remarkable versatility of the enzymes of bacteria and fungi.

The enzyme armament varies considerably in different fungi. Some can attack chiefly the cell membranes, others the cell contents. The *Peronosporaceae, Erysiphaceae,* and rusts, for instance, have little or no action on the membranes, using chiefly some of the cell contents, while those wood-decaying fungi that affect tissues with scanty cell contents must get much of their food from the walls; this accounts for the cavities in the older wood caused by certain species. Enzyme production may vary according to the strain of the fungus concerned, as is fully recognised in brewing and other fermentation industries. The conditions under which the fungus grows, especially its nutrition and aeration, also have a marked effect.

Thus a strain of Aspergillus niger was found to form diastase copiously when grown in media containing starch, glucose, or maltose but not at all when grown on 5 per cent glycerine. An acid reaction was necessary for optimum production on the enzyme in this case, but pectinase may sometimes be most active in acid, sometimes in alkaline plant juices, and this seems to depend on the presence of other substances which are adsorbed from the medium. It is believed that the pectinase is the same in all of a number of fungi tested, but its properties depend on its exposure to the action of other substances, while whether it is produced at all and the quantity produced, depend on the medium in which the fungus is grown.

Some fungi can cause notable rises in temperature in decomposing plant material. Thus Aspergillus fumigatus has been found to raise the temperature of oat straw inoculated with it from 25° to nearly 55° C., in 38 hours. The most extensive decomposition of straw has been obtained by a mixture of soil Actinomycetes and fungi. Most of the common soil moulds Aspergillus, Penicillium, Trichoderma, Rhizopus, and so forth utilise the carbohydrates cellulose, hemicellulose, pentosans, and mannan readily and rapidly in decomposing plant material but not lignin; certain Basidiomycetes! however, such as the common edible mushroom, do not make use of hemicelluloses but deplete the lignin and protein, while causing also a loss of cellulose. Some of these fungi can decompose humus.

Index

A

Abscisic 6
Agrobacterium 87, 90, 91, 92, 93, 94, 95, 96, 98, 99, 101, 102, 103
Agrobacterium-mediated 152, 153, 154, 241
Allele 250, 251, 252
Amylosplast 111
Arabidopsis 36, 57
Aseptic 52
Autotetraploid 254

B

Bacteria 264, 266, 268, 269, 276, 277
Biodiversity 247, 249
Biologists 79
Bioprospecting 246, 247
Bombarded 108
Bombardment 241, 244, 245, 246
Brassicas 254

C

Callus 9, 10, 11, 12, 17, 18, 38, 65, 68, 69
Cell Contents 275, 277, 278
Cell Expansion 76, 77, 104
Cell Growth 104
Cell Membranes 277
Cell Wall 104, 105
Centrosome 112
Cereal Species 241, 244, 245
Chloroplasts 20, 111
Chromosomes 236, 237, 239, 240, 256, 257
Cloning Genes 146, 148
Cloning Plant 146
Crown Gall 264, 267, 268, 269
Cryptochrome 82
Cytoplasm 21, 23, 108, 109, 110

D

Diploid Genotypes 24
Diversity 147
DNA 140, 141, 142, 143
Druse Crystal 112

E

Elimination 9
Elongation 73
Embryo 5, 8, 77
Enzymatic 113, 119, 121
Equivalent 44
Evolutionary 77
Exact Mechanism 84, 108

F

Fermentation 276, 278
Fungi 45
Fungi Enzyme 276

G

Gene Expression 59
Gene Mapping 235
Gene Technology 246, 248

Genetic Analysis 151, 235
Genetic Engineering 145, 147, 241
Genetic Map 151, 237, 240, 249
Genetic Variation 258
Genome 152, 153, 236, 239, 240, 242, 243, 244, 254, 258
Genotype 153, 241, 242, 244, 250, 252, 253, 260, 261
Germination 107
Germplasm 247
Golgi Vesicles 112

H

Haustoria 274, 275, 276
Heteroxylans 114, 117, 118, 119, 122, 124
Hybridization 242
Hydrolysis 121, 124

I

Immigration 250
Inhibition 88, 95
Inoculation 9, 143
Isolation 40, 64, 65, 70

M

Media Components 2
Mitochondria 111
Molecular Markers 254
Monocotyledons 25
Mucilage 31
Multienzyme 124
Mycelium 264, 271, 274, 275, 276, 277

P

Parenchyma Cells 31
Peroxisomes 112
Phenotype 145, 148, 150, 152, 241, 260, 261
Photoperiodism 82, 83
Phytochrome 82
Plant Development 85
Plant Genetic 151
Plant Hormones 15, 83, 259
Plant Materials 15
Plant Traits 146
Plastids 110
Poaceae 114, 115, 120, 122, 125
Pollen 250
Polymerase Chain 151
Proliferation 88
Propagation 11, 12, 13, 69, 139, 143, 144
Protein Structure 146
Protoplasm 105
Protoplasmic 72
Protoplasts 241, 243, 244, 245, 246

R

Recombination 151, 153, 236, 237, 238, 239, 240, 249, 254
Regeneration 19, 20, 23, 32, 33, 103
Reproducible 23, 33
Reproduction 250
Rhizosphere 44
Ribosomes 111

S

Sclerenchyma 137
Self-fertilization 250
Somaclonal 10
Sorghum 153, 242, 244
Spores 271
Strategy 150, 152, 154, 248

T

Tagging 148, 149, 150, 151, 153, 154
Tissue 145, 146, 240, 241, 244, 245, 246, 256, 257, 259
Tissue Elements 54
Transposon 148, 149, 150, 151, 153
Triploid Genotypes 23
Tumefaciens 38

V

Vacuole 30, 31, 110, 111, 112, 133, 134
Vacuoles 12, 133
Variation 231, 244, 246, 250, 251, 252, 253, 258, 259, 260, 261
Vectors 102
Virus 263, 264, 271, 272, 273
Virus diseases 272

X

Xylem 76